VOLUME ONE HUNDRED AND THIRTY

ADVANCES IN
COMPUTERS

VOLUME ONE HUNDRED AND THIRTY

ADVANCES IN
COMPUTERS

Edited by

ALI R. HURSON
*Missouri University of Science and Technology,
Rolla, MO, United States*

ACADEMIC PRESS
An imprint of Elsevier

Academic Press is an imprint of Elsevier
50 Hampshire Street, 5th Floor, Cambridge, MA 02139, United States
525 B Street, Suite 1650, San Diego, CA 92101, United States
The Boulevard, Langford Lane, Kidlington, Oxford OX5 1GB, United Kingdom
125 London Wall, London, EC2Y 5AS, United Kingdom

First edition 2023

ISBN: 978-0-443-19296-8
ISSN: 0065-2458

For information on all Academic Press publications
visit our website at https://www.elsevier.com/books-and-journals

Publisher: Zoe Kruze
Developmental Editor: Nadia Santos
Production Project Manager: Sudharshini Renganathan
Cover Designer: Mark Rogers

Typeset by STRAIVE, India

Working together
to grow libraries in
developing countries

www.elsevier.com • www.bookaid.org

Contents

Contributors

Zeinab Abdalla
Department of Computer Science, University of Denver, Denver, CO, United States

Ahmed Alhaddad
Department of Computer Science, University of Denver, Denver, CO, United States

Anneliese Andrews
Department of Computer Science, University of Denver, Denver, CO, United States

Curtis Brinker
Missouri University of Science and Technology, Rolla, MO, United States

Weibing Chen
College of Electronic Communication & Electrical Engineering, Changsha University, Changsha, China

Tim Guertin
Computer Engineering, Missouri University of Science and Technology, Rolla, MO, United States

Ali Hurson
Department of Electrical and Computer Engineering, Missouri University of Science and Technology, Rolla, MO, United States

Syed Asad Hussain
Department of Computer Science, COMSATS University Islamabad (CUI), Lahore, Pakistan

Elanor Jackson
Missouri University of Science and Technology, Rolla, MO, United States

Justin L. King
Missouri University of Science and Technology, Rolla, MO, United States

Siwei Li
School of Computer Science and Electrical Engineering, Hunan University, Changsha, China

Muhammad Ziad Nayyer
Department of Computer Science, GIFT University, Gujranwala; Department of Computer Science, COMSATS University Islamabad (CUI), Lahore, Pakistan

Imran Raza
Department of Computer Science, COMSATS University Islamabad (CUI), Lahore, Pakistan

Sahra Sedigh Sarvestani
Missouri University of Science and Technology, Rolla, MO, United States

Ming Xia
School of Computer Science and Electrical Engineering, Hunan University, Changsha, China

Gaobo Yang
School of Computer Science and Electrical Engineering, Hunan University, Changsha, China

Preface

Traditionally, *Advances in Computers*, the oldest series to chronicle the rapid evolution of computing, annually publishes several volumes, each one typically comprising four to eight chapters, describing the latest developments in the theory and applications of computing.

The 130th volume is an eclectic one that has been inspired by the recent interest in research and development in computer science and computer engineering. This volume is a collection of five chapters as follows:

In response to an increasing number of app downloads on mobile devices, Alhaddad et al. report and articulate their experiences of developing a black box, model-based testing approach to test mobile apps. Chapter 1, "FSMApp: Testing mobile apps," introduces the FSMApp approach and compares it with the black box MBT approach. Several case studies are presented to explore the applicability, scalability, effectiveness, and efficiency of FSMApp.

An overview of an autonomous vehicle's communication ecosystem within the framework of an intelligent transportation system is the major theme of Chapter 2 entitled "Wheel tracks, rutting a new Oregon Trail: A survey of autonomous vehicle cybersecurity and survivability analysis research." In the chapter, King et al. address the importance of cybersecurity in the development of autonomous vehicles. Vulnerabilities involving existing vehicular technology and connectivity among vehicles at varying levels of autonomy are discussed and recognized as key issues in the survivability of autonomous vehicles under cybersecurity attack. The chapter surveys the research landscape of autonomous vehicles, focusing on security and survivability; related attributes such as performability are also addressed. Research areas are visualized in a taxonomy, and gaps are discussed throughout the chapter. Finally, recommendations and future research opportunities are articulated.

Chapter 3 entitled "ClPyZ: A testbed for cloudlet federation" by Nayyar et al. describe the challenges of finding a suitable instrument and environment to conduct and validate research. The chapter presents ClPyZ, a visualization platform of cloudlet computing, which is a variant of mobile edge computing that aims to provide computational facility to users to enhance the quality of services and the quality of experience for resource-constrained mobile devices. In addition, it explores the concept of federation resource sharing and load balancing.

"The multicore architecture" by Guertin and Hurson is the subject of discussion in Chapter 4. The chapter articulates the importance of a multicore approach in supporting both instruction-level parallelism and thread-level parallelism at a lower clock frequency in comparison to the superscalar approach that offers a higher performance per watt.

However, the advantages of a multicore architecture come at the expense of several challenges such as cache coherency and communication among the cores. The chapter is intended to address these architectural challenges and their potential solutions within the scope of the multicore architecture.

Finally, in Chapter 5 entitled "Perceptual image hashing using rotation invariant uniform local binary patterns and color feature," Xia et al. focus on the perceptual image hashing technique commonly used in cloud-based multimedia systems for security purposes. A four-step scheme is proposed. Extensive experiments are performed to show the robustness and discrimination capability of the proposed scheme and its superiority over several other existing methods.

I hope that the readers will find this volume interesting and useful for teaching, research, and other professional activities. I welcome feedback on this volume as well as suggestions for topics of future volumes.

ALI R. HURSON

Missouri University of Science and Technology

Rolla, MO, United States

CHAPTER ONE

FSMApp: Testing mobile apps

Ahmed Alhaddad, Anneliese Andrews, and Zeinab Abdalla
Department of Computer Science, University of Denver, Denver, CO, United States

Contents

Abstract

A mobile application is a software program that runs on mobile devices. In 2017, 178.1 billion mobile apps were downloaded, and the number is expected to grow to 258.2 billion app downloads in 2022. The number of apps poses a challenge for mobile application testers to find the right approach to test apps. This paper presents a black-box, model-based testing approach to test mobile apps (FSMApp). It is an extension of an

Advances in Computers, Volume 130
ISSN 0065-2458
https://doi.org/10.1016/bs.adcom.2022.09.001

existing approach to test web applications. We present the FSMApp approach and compare the approach with another black-box MBT approach. A number of case studies explore applicability, scalability, effectiveness, and efficiency of FSMApp with this approach. (number of words is 16,349).

1. Introduction

A Mobile Application, or App, refers to software run on mobile phones or smart devices. Millions of Apps are available via App stores like Google Play[a] and Apple App[b] Store [1]. App downloads are projected to increase to 258.2 billion in 2022 from 178.1 billion in 2017 [2]. The pervasiveness of App use also means that quality becomes a major concern. Fierce competition [3] means that a reliable App will be more successful. While Apps share common technology with other software, especially web applications, they differ from desktop software in some important ways [4]: interaction with other applications; sensor handling such as touch screens and cameras; both native and mobile web applications; a multitude of hardware devices and platforms; heightened security concerns; usability that is influenced by other Apps and by the common small size of the smart phone; power consumption; and complexity of testing.

The complexity of testing arises from the fact that, in addition to the same issues as found in web applications, App testing must deal with issues related to mobility, transmission through software, and the issues listed above. Testing mobile Apps is clearly more complex than testing desktop applications [5]. Muccini et al. [6] investigated how mobile App testing differs from testing traditional applications. Mobile connectivity needs to be tested for different connectivity scenarios, networks, resource usage and associated performance degradation possibly resulting in incorrect system functioning. All of these items need to be evaluated, as does energy consumption. Varying device screen resolutions, dimensions, etc., affect usability requiring usability testing. The large combination of platforms, operating systems, diversity of devices, and rapid evolution is challenging for a tester, as it can lead to test explosion. Performance assessment is crucial. Many of these testing needs require that a functional test be executed for a number of specific environmental scenarios, set-ups, and devices.

---

[a] https://play.google.com/store?lil=en.
[b] https://itunes.apple.com/us/genre/ios/id36?mt=8.

This is one reason why test automation is clearly desirable and has been pursued quite successfully [7–15]. In most cases, the tools are not based on a model-based testing approach and still require the development of a test suite up front. They capture test inputs and play them back, or simply automate existing tests for different configurations, devices, and platforms.

Our interest is in the model-based black–box testing of mobile applications. Specifically, we are interested in extending an existing technique, FSMWeb [16] to test mobile Apps. FSMWeb [16–19] is a widely cited approach that tests web applications. Andrews et al. [16] proposed FSMWeb as a black-box model-based testing approach. The model consists of a hierarchical collection of FSMs. In addition, Andrews et al. [20] study the scalability issues of traditional FSMs of web applications compared to FSMWeb. FSMWeb compresses inputs using a special purpose input constraint language [20] reducing the model by as much as 90% . The case studies [20] show that FSMWeb is more efficient than conventional FSM techniques. Ran et al. [21] defined input selection for FSMWeb. Andrews et al. [17,22] also propose an approach for selective regression testing of web application using FSMWeb and develop a cost-benefit trade-off framework between brute force and selective regression testing.

This paper is organized as follows: Section 2 describes existing work related to black–box testing approaches in testing mobile apps, and the original FSMWeb approach. Section 3 presents the extensions to the FSMWeb approach for testing mobile application (FSMApp). Section 4 compares FSMApp with another approaches for MBT for mobile apps [23] using one small example app. Section 5 describes a number of case studies to compare FSMapp this other approach and explores applicability, scalability, effectiveness, and efficiency. Section 6 draws conclusions and suggests further work.

2. Background

The background section first summarizes Black-Box Model-Based Testing (MBT) techniques to test web applications; then it explores existing work for Black-Box MBT approaches for testing mobile Apps.

2.1 Black-box model-based testing

Nguyen et al. [24] define Model-Based Testing (MBT) as an approach to generate test cases using an abstraction of the system under test (SUT). The model provides an abstract view of the SUT by focusing on particular system characteristics. Utting et al. [25] provide a survey on MBT. They define six dimensions of MBT approaches: model scope, characteristics, paradigm, test selection criteria, test generation technology, and test execution. Dias-Neto et al. [26] characterize 219 MBT techniques after analyzing 271 MBT papers and describe approaches that support the selection of MBT techniques for software projects, including risk factors. Risk factors may influence the use of these techniques in industry. Utting et al. [25] classify MBT by notation used, such as State-Based, History-Based, Functional, Operational, Stochastic, and Transition-Based. Transition-Based notations are graphical node-and-arc notations that focus on defining the transitions between states of the system, such as Finite State Machines (FSMs). Transition-Based notations also include UML behavioral models, such as activity diagrams, sequence, and interaction diagrams [25]. This paper addresses the use of MBT for functional testing of mobile applications. We apply MBT and utilize a hierarchical finite state machine model as described in Section 3.

Researchers provide many MBT techniques for web applications such as [16,27–34]. Andrews et al. [16] propose an approach to test web applications with Finite State Machines (FSMWeb). The approach is based on a black box functional model. FSMWeb is a hierarchical collection of FSMs. The approach consists of two phases: (1) building a model of the web application and (2) generating a test suite from the model. The first phase is completed in four steps: an application is divided into clusters, logical web pages are defined, FSMs are built for each cluster, and for the application (top) level [16]. The second phase is completed in three steps: (1) test paths for each cluster are generated by a variety of coverage criteria, such as edge coverage, (2) path aggregation generates abstract test paths, and (3) inputs are selected for the abstract test paths.

2.2 Testing mobile apps

Our main interest in testing Mobile Apps is Black-Box functional testing of mobile apps. As such, we are not interested in other testing activities for Mobile Apps, such as usability testing, configuration testing, exception handling testing, etc. This is reflected in how we review the literature. We first survey approaches for MBT for testing Mobile Apps. Then, we quickly review major approaches for test automation, primarily to determine what options we have to turn the tests generated by an MBT approach into executable tests.

2.2.1 MBT and App testing

Sahinoglu et al. [35] present a mapping study of testing mobile applications. Their paper studies the research issues in mobile application testing and the most frequent test type and test level of available studies in mobile testing. One of the categories relates to MBT of Apps. There are only six model-based testing studies out of 123 studied. This lack of MBT techniques shows that more work is needed. Here, we focus on the use of a behavioral model such as a state transition diagram to generate tests. Another systematic mapping study by Méndez-Porras et al. [9] discusses 83 empirical studies of automated testing of mobile applications. The categories include Model-based testing, Capture/replay, Model-learning, Systematic testing, Fuzz testing, Random testing, and Script based testing. This paper focuses on Black-Box MBT of mobile applications. Hence, only the first category is relevant to the scope of this paper. Méndez-Porras et al. [9] cite [10,13,15,36–44] as model-based testing papers. We exclude a number of papers from this list as they are not in our scope. This subsection does not discuss white-box testing [13,38] and grey-box testing [10]. We are also not interested in testing security [40], data-flow analysis [42] and life cycle testing of the application [44]. Zaeem et al. [37] is not discussed in this paper because it is related to oracle problems. Wang et al. [15] identify several difficulties for automating GUI testing and study the high cost for achieving coverage of the traversal algorithm to generate a GUI tree model. It extends the crawling technique in Amalfitano et al. [36] to increase GUI coverage. Amalfitano et al. [39] present a Event-Based Testing approach that Méndez-Porras et al. [9] classify as MBT. This paper focuses on creating the model for testing mobile application and we discuss Amalfitano et al. [36] later as one of the approaches that we will compare to FSMApp.

Several research papers are not included in Méndez-Porras et al. [9]. Zein et al. [45] present anther mapping study of mobile application testing techniques. The goal of the mapping study is based on the classification of empirical studies. The main research question is what are the empirical studies that investigate mobile application testing techniques and what are the challenges? Zein et al. [45] consider studies of mobile testing techniques, services,[c] security and usability testing of mobile applications, and the challenges of testing mobile applications. The mapping study uses five categories: Usability testing, Test automation, Context-awareness, Security, and a

---

[c] "Mobile services are currently targeting time and safety critical contexts such as abnormal and disaster management situations" [45].

general category. The general category includes all studies which are not in the other areas. None of the categories defined in this mapping study directly speaks to the scope of this paper. However, the category Test Automation contains two potentially relevant subcategories: Model-based test automation and Black-box test automation. The other categories are not relevant to our scope. Zein et al. [45] classify [8,11,23,36,46–49] as Model-based testing methods. Not all are relevant to this paper. We exclude [8] because it is a combination of model-learning and model-based testing and we are interested in model-based testing only. Also, Ref. [48] focuses on testing environments such as connection to the wireless network. Ref. [49] is a white-box testing approach.

Examining the papers listed under Model-based testing and Black-box testing yielded the following papers that fall into our scope [8,11,23, 36,46–50]. The Model-based testing approach builds a model of the application being tested and uses this model to generate tests. We consider four types of MBT: (1) The user will create the model manually, (2) The tool will generate the model automatically, (3) The user generates the test manually, (4) The tools generate tests and execute them.

We summarize research that uses manual or automated model building or test generation. [23,41,46] build the model manually and [8,11,36, 47,49–51] build the model automatically. The manual model is [23,41]. All methods generate the test cases automatically for at least part of test generation process.

This paper compares FSMApp with Refs. [23,36] because they describe the generation of the model in sufficient detail. First, we will describe the MBT methods using manual model generation, then those that use an automated method.

Jing et al. [41] present a Model-based Conformance Testing Framework (MCTF) for Android Applications. Testers need to generate the model manually from the requirements. MCTF consists of four steps: System Modeling, Test Case Generation, Test Case Translation, and Test Case Execution. System Modeling derives the parameters and properties of an App, while Test Case Generation generates abstract tests automatically with Alloy Analyzer. Alloy Analyzer is a language for describing structures and exploring them. Test Case Translation converts abstract tests into executable tests. Test Case Execution compiles executable test cases to generate test packages. The Android apps or operating system is then tested with the packages. Packages are defined as test cases generated by Alloy Analyzer. The packages are run by Android's Instrumentation Test Runner. They can access the Android's applications.

Lu et al. [46] propose an activity page based model to automate functional testing for mobile applications. The Activity page based model is a directed graph. An activity page is similar to a screen and modeled as a tuple that appears to describe input constraints and related actions. Edges represent a trigger event between the states. The model can be generated either by (1) the UIAutomation tool [52] or (2) Create a Model based on an image comparison for every event. The Monkeyrunner tool [53] takes the screenshot of the activity page with the fired event. They generate test cases by applying a crawling algorithm. Lu et al. [46] present two modeling methods based on an activity page based model.[d] However, it is hard to compare FSMApp with this activity page based model because (1) the description of the input constraints is incomplete, especially related to dependencies between inputs and (2) it is unclear at which phase the input constraints are resolved into values. When we used the crawling algorithm to generate test paths, it also became clear that for the small example in Section 4, the paths were inordinately large.[e]

Amalfitano et al. [11] present a GUI automated technique to test Android apps. They implement the technique in a tool called AndroidRipper. The technique is based on a GUIRipper. This is a dynamic approach in which the software's GUI is automatically traversed by opening all its windows and extracting all their widgets (GUI Objects), properties, and values [54]. GUIRipping generates a tree model. The tool takes a long time to generate test cases for large mobile apps, and it does not support some inputs, such as sensors, whereas FSMApp tests large apps and includes more input types.

Costa et al. [47] adopt Pattern-Based GUI Testing (PBGT) to test mobile apps based on User Interface Test Patterns [55] specifically developed to test web applications. This approach does not include some gestures and components like swipe and zooming whereas FSMApp (Section 3) supports components. It does not support varying screen sizes and loops. It needs a separate model for every function of the mobile app. They cannot be connected together to create a hierarchical model to test the mobile app. Costa et al. [47] use a domain specific language to model and test mobile applications. Because their annotation does not use a node and edge notation like FSMApp, it is difficult to compare the two approaches. Therefore, we decided not to include this method in our case study comparison.

---

[d] We focus on the activity page based model from GUI ignoring the method that uses the source code because our goal is black-box functional testing.
[e] 62 paths consisting of 350 nodes, compared to FSMApp 4 paths with 45 nodes.

Takala et al. [51] present a test automation solution for testing Android apps. They use Model-based testing tools (TEMA) [51]. The TEMA tool is a set of Model-based tools for different phases of MBT. The phases are modeling, design, generation and debugging tests. TEMA models are Labeled State Transition Systems (LSTS) [56,57] which contain: state, transition, actions, and labels. The model can be divided into small components that are connected. Each component has two levels: an action machine and a refinement machine. The action machine explains the high-level functionality of the apps with words (action of transition) and state verification (state of SUT). The refinement machine describes action words and state verification using a keyword. The keyword is an abstraction of a user action or state verification such as "press key" or "search text." The model is complex even for small apps because each component is a combination of action machine and refinement machine of each function of a mobile application. The two levels make testing of large mobile apps difficult because a refinement machine state model can only connect to one action machine. In our approach, we implement hierarchical levels, and the clusters can be used for many states which mitigate the state explosion problem and simplify the generation of a model for large mobile apps.

Takala et al. [51] tested the BBC News Widget. It took a few days to generate the model using the TEMA tool. The tool generated 240 test cases with 27,000 action words and 50,000 keywords. The total time to execute the test cases was 115 h with an average of about 30 min for each test case. We applied the approach to the todo app [58] because it is a small app. We created the model manually because the TEMA tool does not have clear documentation for installation and use. The model of the todo app [58] has 37 nodes, 61 edges, 1700 keywords, and 900 action keywords. We estimated that it would take 4 h to run the test cases with the keywords. The number of test paths of this approach compared to FSMApp is very large because FSMApp only required 11 min for executing all test paths. It is complicated to generate the model and the keywords manually. For these reasons, we excluded this approach from our comparison.

Baek et al. [50] present a Model-based Black-box approach to test Android apps with a set of multi-level GUI Comparison Criteria (GUICC). The approach creates a directed graph which contains ScreenNodes and EventEdges. The ScreenNode is a GUI state which includes screen information. GUI state represents the type of GUI information: package name, activity name, layout and executable of the widgets, and content information. The framework has three modules: (1) The communication layer connects between desktop and mobile device. (2) The EventAgent is a tool to run the test on a mobile device. (3) The testing Engine generates a GUI graph with GUICC, test inputs, and test cases. The framework is not open to the public and Baek et al. [50] do not explain the approach in enough detail. Therefore we are not able to compare the approach with our approach.

Amalfitano et al. [36] present a technique and a tool for crash testing and regression testing for Android apps. Amalfitano et al. [36] use a crawler technique to automatically build the model from the GUI and generate the test cases automatically. The GUI model is a tree and hence does not allow for testing loops. The GUI model is a result of testing, while FSMApp builds the model first and then generates tests.

de Cleva Farto et al. [23] evaluate the use of MBT to verify and validate mobile applications through automated tests. They use an Event Sequence Graph (ESG) to build a test model of the App under test. Section 4 describes and compares this approach with FSMApp in detail with an example.

2.3 FSMWeb approach

FSMWeb [16] consists of two major phases: (1) Build a hierarchical model *HFSM*, and (2) Generate tests from the *HFSM*.

The term *cluster* is used to refer to collections of software modules/web pages that implement a logical user level function. The first step partitions the web application into clusters. At the highest level of abstraction, clusters represent functions that can be identified by users. At a lower level, clusters represent cohesive software modules/web pages that work together to implement a portion of a user level function.

Web pages are modeled as multiple *Logical Web Pages* (LWPs). A LWP is either a physical web page or the portion of a web page that accepts data from the user through an HTML form and then sends the data to a specific software module. LWPs are described in terms of their sets of *inputs* and *actions*. All inputs in a LWP are considered atomic: data entered into a text field is considered to be only one user input symbol, regardless of how many characters are entered into the field. There may be rules about the inputs. Some inputs may be required; others may be optional; users may be allowed to enter inputs in any order, or a specific order may be required. Table 1 shows how typical input types found in web applications are represented as constraints on edges in an FSMWeb model.

The lowest level cluster FSMs are generated with only LWPs and navigation between them. Input-action constraints annotate each edge [16].

Table 1 FSMWeb constraint of typical input types.

Input type	FSMWeb edge annotation
Text Field Text Area Field	R (input name)
Optional Text Field Optional Text Area Field Optional Checkbox	O (input name)
Radio Box Drop Down Box (with n options)	C1 (option 1, …, option n)
Optional Radio Box (with n options)	O (C1 (option 1, …, option n))
Set of Checkboxes Multi-Select Box (with n options requiring 0 to n selections)	O (Cn (option 1, …, option n)) A (option 1, …, option n)

Higher level FSMs represent FSMs from a lower level cluster by a single node and may contain LWP nodes as well. Ultimately, a top-level Aggregate Finite State Machine (AFSM) is formed and represents a finite state model of the application at the highest level of abstraction. Note that building the model is possible very early in the development process and need not be a reconstructing activity after the web application is implemented. Software developers have the flexibility to choose the model they consider best. Test sequences are generated during phase 2 of the FSMWeb method. A test sequence is a sequence of transitions through the application FSM and through each lower level FSM. FSMWeb's test generation method first generates paths through each FSM based on some graph coverage criterion such as *edge coverage*. These paths are then aggregated based on an aggregation criterion for each FSM's paths, such as *all combinations* or *each path at least once* [16].

This process results in a set of aggregate paths. We call them *abstract tests*. The final step of the test generation is selecting inputs to replace the input constraints for the transitions of the aggregate paths.

Input selection uses a technique [21] that builds two databases: a *synthetic database*, which consists of values that are consumed during testing, and an *application database*, which contains values previously inserted by the application being tested. Values are saved into the application database during execution and saved into the synthetic database during testing. Details about the database creation and input selection can be found elsewhere [21].

3. Extensions: Testing mobile apps

3.1 Testing process for mobile apps

Our approach to testing Mobile Applications using Finite State Machines (FSMApp) is an extension of FSMWeb [16] which is described in this section. Fig. 1 shows the phases of the FSMApp process. FSMApp proceeds in four phases (FSMWeb has only three): Phase 1 builds a hierarchical model HFSM, Phase 2 generates tests from the HFSM, Phase 3 selects the inputs and Phase 4 compiles and executes tests through automated mobile testing tools.

3.2 Example used to illustrate approach

We illustrate our approach using the Family Medicines List application as the example. The Family Medicines List [59] is an open source mobile application offering the essential functions to manage medical information for a family. It is built using Basic4Android (B4A) [60] for Android operating systems. B4A is a graphical user interface tool to create native Android

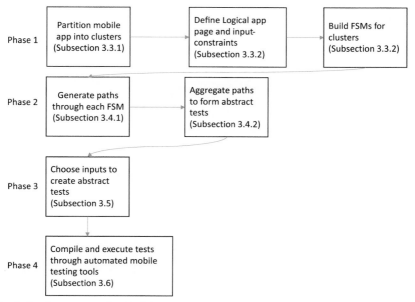

Fig. 1 Testing process for Mobile Apps.

applications where the backend of the app is programmed in Java. Fig. 2 shows screens for the Family Medicines List.

The app has the following basic functionalities: (1) manage medicine information for a set of medicines. This includes adding, editing, deleting, and searching functions, (2) manage family member information alongside their medicine, (3) manage dosage and individual instruction for each medicine, and (4) list view with images for medicine lists. Fig. 2 shows the three main pages of the Family Medicines List. Fig. 2A shows the list of all the medical information in the view list (component) of a family member after selecting a name from the list of users, Fig. 2B displays the form for user information management based on a database of all the information the user entered, and Fig. 2C add information about new medication page.

3.3 Phase 1: Build model
3.3.1 Partition the mobile app into clusters (Cs)
The term *cluster* is used to refer to collections of software modules/app pages that implement a logical or user level function. The first step partitions the app into clusters. At the highest level of abstraction, $i = 0$, clusters represent functions that can be identified by users. Hence, an $HFSM = \{FSM_i\}^n$ with a top level $FSM_0 = AFSM$. Each FSM has nodes that represent either Logical App Pages (LAPs) or clusters. Edges are internal or external to an FSM. External nodes span cluster boundaries. (They become internal at the next higher level.) External edges can either enter or leave a cluster FSM.

Fig. 2 Family Medicines List App.

Clusters may be an individual Activity or software modules that represent a major function. Clusters can be identified from the site navigation layout, coupling relationships among the components, and design information. Similar to FSMWeb, the FSMApp model can be constructed early in the software life cycle and we provide the developer the flexibility to create the model they consider best. Our example has one system to manage the medication of each member of a family. The lower level clusters are Info member medication (new name), Add and edit new medication of the family member. Fig. 3 shows the top level of the Family Medicines List App. There are three main clusters: entering a new medication, entering a new name, modifying a medication and exiting the App.

We will use Android mobile application to illustrate our approach. Android is a mobile operating system (OS) based on the Linux kernel [61]. It is designed for touchscreen mobile devices, for example, smartphones and tablets. Mobile applications fall in to three categories: native, web-based or hybrid. Native mobile applications are built to run directly on the OS. Web-based apps run on the browser of the device. A hybrid app is a combination of native app and web-based apps. Obviously since it is a black-box approach, it can be applied to other operating systems such as IOS or Windows with their components.

We apply FSMApp to the Family Medicines List App. Fig. 3 shows the top level of the Family Medicines List App, while Figs. 4–12 show the detail for the other clusters of the Family Medicines List App.

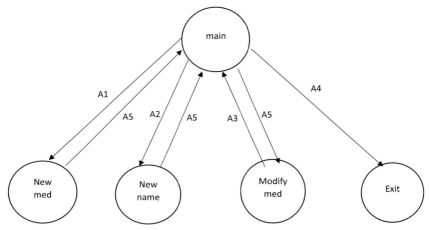

Fig. 3 Main Page.

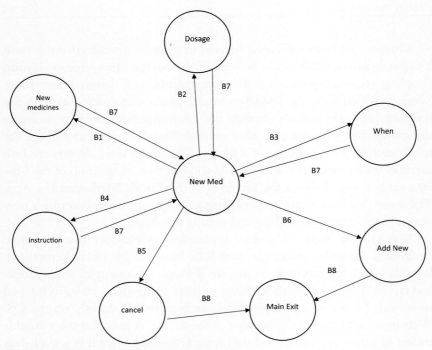

Fig. 4 Family Medicines List New Med.

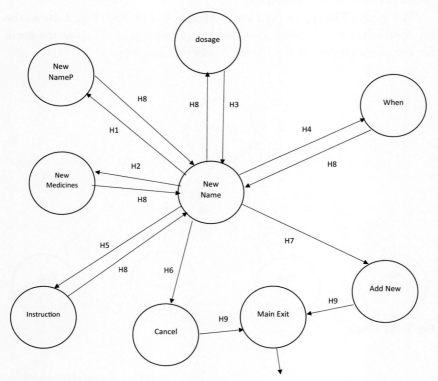

Fig. 5 Family Medicines List New Name.

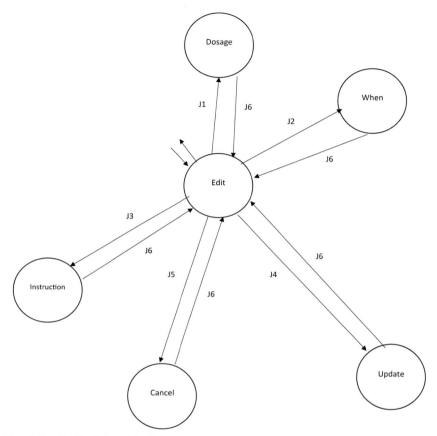

Fig. 6 Family Medicines List Edit.

We classify the main Screen (A) into three clusters (B, C, D) as shown in Fig. 3. The AFSM for this example also the exit of the application as Exit LAP. We describe cluster (B), (C) and (D) in detail. Cluster (B) has four subclusters: New med in Fig. 9, Dosage in Fig. 10, When in Fig. 11 and Instruction in Fig. 12. We have two LAPs with two buttons, cancel and add New, as shown in Fig. 5. Cluster (C) has five subclusters: "New name" in Fig. 8, "New med" in Fig. 9, "Dosage" in Fig. 10, "When" in Fig. 11 and "Instruction" in Fig. 12. The Main Screen has two LAPs with two buttons, cancel and add New, as shown in Fig. 4. Cluster (D) has one subcluster which is Edit in Fig. 6. We have two LAPs with two buttons, delete and cancel, as shown in Fig. 7. Subcluster(E) has three subclusters: Dosage in Fig. 10, When in Fig. 11 and Instruction in Fig. 12. We have two LAPs with

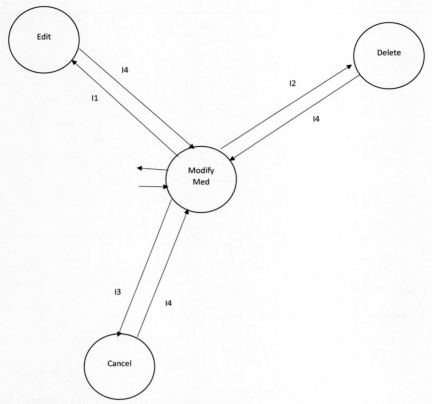

Fig. 7 Family Medicines List Modify Med.

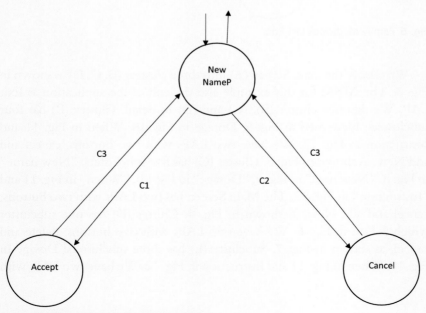

Fig. 8 Family Medicines List New NameP.

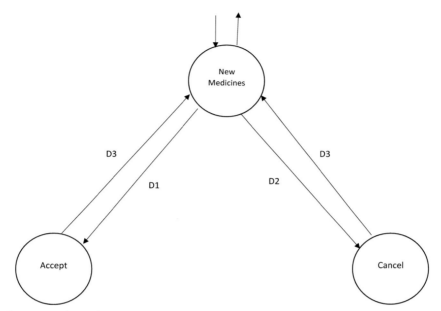

Fig. 9 Family Medicines List New Medicines.

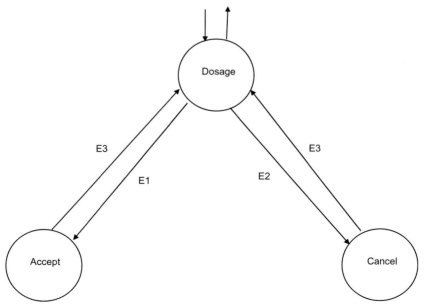

Fig. 10 Family Medicines List Dosage.

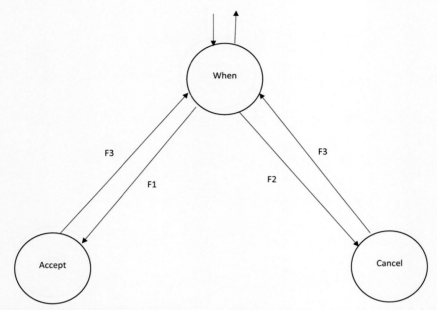

Fig. 11 Family Medicines List When.

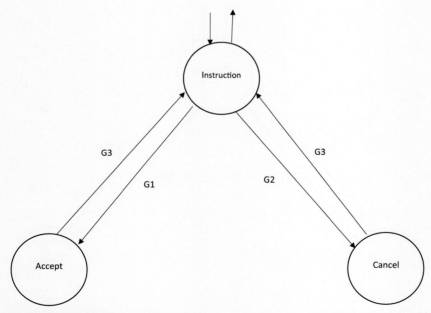

Fig. 12 Family Medicines List Instructions.

two buttons, cancel and add New, as shown in Fig. 6. Also, We have two LAPs with two buttons, cancel and accept, for each subclusters (F, G, H, I, J) shown in Figs. 8–12.

3.3.2 Define logical app pages (LAPs) and input-action constraints for each

Mobile apps have a variety of screens. We will consider screens as input components, or logical app pages (LAPs), and the inputs and their constraints on these LAPs next. For illustration purposes, we list the actives (screens) for Android. Similar components exist for other types of mobile devices. Many app actives (screens) contain XML forms, each of which can be connected to a different back-end software module. To facilitate testing of these modules, app pages are modeled as multiple *Logical App Pages* (LAPs). A LAP is either a physical app page, physical app component, or the portion of an app activity that accepts data from the user through a XML form, and then sends the data to a specific software module. FSMApp is an MBT meant for black-box testing hence, the mobile application can be written in any language appropriate for mobile applications (e.g., Ruby, JavaScript, HTML, etc.). All inputs in a LAP are considered atomic: data entered into a text field is considered to be only one user input symbol, regardless of how many characters are entered into the field.

There may be rules about the inputs: some inputs may be required, while others may be optional; users may be allowed to enter inputs in any order, or a specific order may be required. Required (R) means that required input must be entered. Required Value (R(parm)) means that one must enter at least one value. Optional (O) means that an input may or may not be entered. Single Choice (C1) means that one input should be selected from a set of choices, and Multiple choice means that more than one input should be selected from a set of choices. Table 2 shows how typical input types found in mobile applications are represented as constraints on edges in an FSMApp model. The difference between the web input types and mobile input types are swipe and scroll. Swipe (W) means that a swipe is required to change the value of a component. Scroll (L) indicates that the input required is to scroll up or down the content.

FSMWeb [16] and FSMApp differ in their input types. FSMApp has many more input types than FSMWeb. FSMApp models components, whereas FSMWeb does not provide for components. Mobile applications can have a variety of components that can be modeled via input constraints.

Table 2 Components of Mobile Application (LAPs).

No	Components	Interface controls	Actions	Effect	Input type
1	Bottom sheets	A. Button	A. R(Button, click)		Non–Text
		B. Link	B. R(Button = X, click)	Close	
2	Buttons	A. Floating action button	A. R(Content, click)	Search	Non–Text
		B. Raised button	B. R(Content, click)	Save	
		C. Flat button	C. R(Button, click)		
			D. R(Button, click)	Show list or select button	
3(c)	Cards	A. Image	A. R(Image, click)	Display	Text
		B. Video	B. R(Image, click)	Change size	
		C. Textbox	C. R(Video, click)	Run	
		D. Text Area	D. R(Textbox)	Display text	
		E. Button	E. R(Text Area)	Display many lines of text	
		F. Links	F. R(Button, click)		
			G. R(Link, click)	Show other website or page	
4(c)	Chips	A. Textbox	A. R(Enter text)	Save	Text
		B. Cards	B. R(Display content, choose)	Show details	
			C. R(Click on chip)	Display card	

Table 2 Components of Mobile Application (LAPs).—cont'd

No	Components	Interface controls	Actions	Effect	Input type
5(c)	Data Tables	A. Checkbox	R(Select, select dialog, add content)	Save	Text
		B. Link	R(Select, click button)	Delete	Non–Text
		C. Textbox	R(Click link)	Transfer	
		D. Menu	R(Select)	Show card	
		E. Button			
		F. Card			
6(c)	Dialogs	A. Button	R(Show warning, click close)		Text
		B. Textbox	R(Select dialog, in- put content)	Save	Non–Text
		C. Date picker	R(Select dialog, click date picker, choose date, close)	Save	
		D. Checkbox	R(Select dialog, click Time picker, choose date, close)	Save	
		E. Time picker	R(Click menu, choose from list)	Close	
		F. Radio Box			
		G. Menu			
		H. Bar slide input			

Continued

Table 2 Components of Mobile Application (LAPs).—cont'd

No	Components	Interface controls	Actions	Effect	Input type
7	Dividers	A. Images	A. R(Show Divider)	Show images	Text
			B. R(Show Divider)	Show content	
8	Grid lists	A. Images	A. R(Select image, zoom in)	Show images list	Text
		B. Text	B. R(Select grid list, scrolling)	Show text list	
			C. R(Select title, sort)	Sort text	
9	Lists	A. images	A. R(Select title, sort)	Show title list	Text
		B. Text		Sort title list	
10(c)	Menus	A. Button	A. R(Select text, copy)	Show list in a menus	Text
		B. Text	B. R(Select combobox, choose content)	Links of button to another pages	Non–Text
		C. Combobox	C. R(Select text, write content)		
		D. Checkbox			
		E. Switch			
		F. Reorder			
		G. Expand/collapse			
		H. Leave-behinds			
11(c)	Pickers	A. Dialog	A. R(Select dialog, choose info)	Save	Non–Text
			B. R(Select dialog, cancel)		

Table 2 Components of Mobile Application (LAPs).—cont'd

No	Components	Interface controls	Actions	Effect	Input type
12	Progress & activity	A. Button	A. R(Click button)	Loading	Non-Text
13	Selection controls	A. Checkbox	A. R(Click checkbox, change behavior of page)	Change the behavior of the page	Non-Text
		B. Radio Buttons	B. R(Click radio button, change behavior of page)		
		C. On/Off switches	C. R(Change switch, change behavior of page)		
14	Sliders	A. Slide bar	A. R(Change slider, effect on page)	Insert input Change behavior of the page	Non-Text
15	Snackbars & toasts	A. Button	A. R(Click button, display change)	Dismiss or cancel the action	Text
		B. Text	B. R(Show text)		Non-Text
16	Subheaders	A. Button	A. R(Click button, change page)	Filtering or sorting the content	Text
		B. Text			Non-Text
17	Steppers	A. Button	A. R(Click button = next, next step) B. R(Click button = previous, previous step) C. R(Click button = cancel, cancel process)	Show feedback of the process	Non-Text

Continued

Table 2 Components of Mobile Application (LAPs).—cont'd

No	Components	Interface controls	Actions	Effect	Input type
18	Tabs	A. Dropdown Menu	A. R(Select tab)	Show content	Text
		B. Text label	B. R(Select tab)	Show drop menu	
19	Toolbars	A. Button	B. R(Click toolbars, display list, click button)	Display the list	Non–Text
20	Tooltips	A. Images	A. R(Hover images)	Show text	Text
21	Text fields	A. Single-line text field	A. R(Content, click)	Save the content	Text
		B. Floating Label			
		C. Multi-line text field			
		D. Full-width field text field with Character counter			
		E. Multi-line with character counter			
		F. Full-width text field with character counter			
		G. Auto-complete text field			
		H. Inset auto-complete			
		I. Full-width inline auto-complete			
		J. In-line auto-complete			

While they vary a little between different mobile operating systems, they also have many types of components in common. To illustrate what components look like Table 2 summarizes common components of Android applications and what FSMApp's input constraints would look like.[f] LAPs are at the lowest model level. They can be an input type as defined in Table 2 or a component. For example, "card" is a component. A component can contain another component. Column 1 shows the number of components. The (c) mark means that the component can contain another component. Column 2 of Table 2 shows the name of the component. Column 3 shows the interface controls (i.e., the input types for the mobile application or for the mobile component). Column 4 shows the input constraints and transition information. Column 5 shows the effect of executing the component, when entering inputs that satisfy the input constraint. The last column represents the types of interface control: Text and Non-Text.

Transitions connect the nodes and the clusters. They are annotated with input constraints to indicate what inputs and actions lead to the next node or cluster. Fig. 13 shows the Add New Medicines input action constraints. In the New Medicines cluster, there are two states: either you enter the new medicine by name (parMed) and accept it (buttonANMA), or you cancel the new medicine (with or without giving the medicine name). Incoming and outgoing edges for this cluster connect to the parent cluster. They don't require any user actions. We also added two dummy transitions to keep the graph single-entry-single exit.

Tables 3–12 show the transitions, explanation, and input action constraints. Column 1 uniquely identifies each transition. Column 2 shows an explanation of the transition. Column 3 shows all input action constraints with all required or optional inputs. The corresponding graphs are

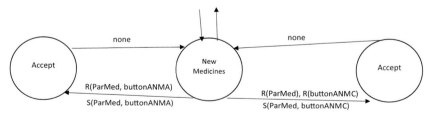

Fig. 13 Annotated FSM for New Medicines cluster of Table 8.

---

[f] These are the 21 components that Google lists for Android Apps. We use google terminology and that component have also be called "widgest."

Table 3 Transitions of Fig. 3 (main page cluster AFSM).

Transition	Explanation	Constraints
A1	Access New Medicines	R(selectN, buttonANM) S(selectN, buttonANM) Continue-use(SelectN)
A2	Access New Name	R(buttonANN)
A3	Access Modify Medicines	O(selectN), R(selectM) S(selectN, selectM) Continue-use(SelectN, SelectM)
A4	Exit the System	R(buttonBack)
A5	Back to Main Page	none

Table 4 Transitions of Fig. 4 (New Med cluster).

Transition	Explanation	Constraints
B1	Access New Medicines	R(buttonANM)
B2	Access Dosage	R(buttonAD)
B3	Access When	R(buttonAW)
B4	Access Instructions	R(buttonAI)
B5	Cancel Add New Med	R(buttonCNM)
B6	Add New Medicines	R(buttonAM)
B7	Cancel to New Med	None
B8	Back to Main Page	None

Table 5 Transitions of Fig. 5 (New Name Cluster).

Transition	Explanation	Constraints
H1	Access New NameP	R(buttonANNP)
H2	Access New Medicines	R(buttonANM)
H3	Access Dosage	R(buttonAD)
H4	Access When	R(buttonAW)
H5	Access Instructions	R(buttonAI)
H6	Cancel Add New Name	R(buttonCNN)
H7	Add New Name	R(buttonANNA)
H8	Cancel to New Name	None
H9	Back to Main Page	None

Table 6 Transitions of Fig. 7 (Modify Med Cluster).

Transition	Explanation	Constraints
I1	Edit the Medicine	O(Parname, parMed), R(buttonE) S(Parname, parMed, buttonE) Continue-use(parName, parMed)
I2	Delete the Medicine	O(Parname, parMed), R(buttonD) S(Parname, parMed, buttonD)
I3	Cancel to Previous Page	O(Parname, parMed), R(buttonCE) S(Parname, parMed, buttonCE)
I4	Back to Previous Page	None

Table 7 Transitions of Fig. 8 (New NameP Cluster).

Transition	Explanation	Constraints
C1	Add the new name	R(parName, buttonANNPA) S(parName, buttonANNPA) Continue-use(parName)
C2	Cancel to Previous Page	O(parName), R(buttonANNPC) S(parName, buttonANNPC)
C3	Back to Previous Page	None

Table 8 Transitions of Fig. 9 (New Medicines cluster).

Transition	Explanation	Constraints
D1	Add the new Med	R(parMed, buttonANMA) S(parMed, buttonANMA) Continue-use(parMed)
D2	Cancel to Previous Page	O(parMed), R(buttonANMC) S(parMed, buttonANMA)
D3	Back to Previous Page	None

Table 9 Transitions of Fig. 10 (Dosage Cluster).

Transition	Explanation	Constraints
E1	Add the new dosage	R(parDosage, buttonADA) S(parDosage, buttonADA) Continue-use(parDosage)
E2	Cancel to Previous Page	O(parDosage), R(buttonADC) S(parDosage, buttonADC)
E3	Back to Previous Page	None

Table 10 Transitions of Fig. 11 (When Cluster).

Transition	Explanation	Constraints
F1	Add the new when	R(parWhen, buttonAWA) S(parWhen, buttonAWA) Continue-use(parWhen)
F2	Cancel to Previous Page	O(parWhen), R(buttonAWC) S(parWhen, buttonAWC)
F3	Back to Previous Page	None

Table 11 Transitions of Fig. 12 (Instructions Cluster).

Transition	Explanation	Constraints
G1	Add the new instructions	R(parInstructions, buttonAIA) S(parInstructions, buttonAIA) Continue-use(parInstructions)
G2	Cancel to Previous Page	O(parInstructions), R(buttonAIC) S(parInstructions, buttonAIC)
G3	Back to Previous Page	None

Table 12 Transitions for Fig. 6 (Edit Cluster).

Transition	Explanation	Constraints
J1	Access Dosage	R(buttonAD)
J2	Access When	R(buttonAW)
J3	Access Instructions	R(buttonAI)
J4	Update the Medicine	R(buttonEEU)
J5	Cancel to Previous Page	R(buttonEEC)
J6	Back to Previous Page	None

mentioned in the caption of the tables. Table 3 shows 5 transitions for the main page cluster (AFSM). The transitions connect the main page with four clusters and one LAP (Exit App). Table 4 shows 8 transitions for New Med cluster to connect with 4 clusters and two LAP nodes (Update and Cancel). Table 5 shows 9 transitions for New cluster to connect with 5 clusters and 2 LAP nodes (Update and Cancel). Table 6 shows 4 transitions for modify med cluster to connect with Edit medicine cluster, delete LAP and cancel LAP.

Table 7 shows 3 transitions for New NameP cluster to connect with accept LAP and cancel LAP. Table 8 shows 3 transitions for New Medicines cluster to connect with accept LAP and cancel LAP. Table 9 shows 3 transitions for Dosage cluster to connect with accept LAP and cancel LAP. Table 10 shows 3 transitions for When cluster to connect with accept LAP and cancel LAP. Table 11 shows 3 transitions for Instructions cluster to connect with accept LAP and cancel LAP. Table 12 shows 6 transitions for Edit cluster to connect with 3 clusters (Dosage, When, Instructions) and two LAP nodes (Update and Cancel).

In addition to input-action constraints, there may be rules how and whether selected input values may be or must be reused (propagated). The types of the propagated input values are:Continue-use: the selected input values must be reused later in the test path. For example, the patient name must be passed to add new medicines cluster.

- Single-use: the selected input value must be used only once. For example, when one deletes the medicine of a patient, it cannot used in the test again, unless a new medicine is added with this name.
- Not-propagated: The input has no constraints on reuse. We may or may not use it later in the test.

Table 3 shows the input constraints and the propagation rules for the main page (AFSM) of Fig. 3.

Column 1 shows the Transition Annotation. Column 3 shows the input constraints.

3.4 Phase 2: Generate test sequences

3.4.1 Paths through FSMs/AFSM

Test sequences are generated during phase 2 of the FSMApp method. The user can select coverage criteria such as node, edge, edge-pair, simple round trip and prime path coverage. A test sequence is a sequence of transitions through the aggregate FSM and through each lower level FSM. FSMApp's test generation method first generates paths through each FSM based on some graph coverage criterion, such as *edge coverage*.

We generate the test sequences for each cluster that satisfy edge coverage. Table 13 shows test paths for each cluster as sequences of nodes. The corresponding graphs are mentioned in the caption. Nodes in bold indicate the node is a cluster node. For this App, with 10 clusters, several clusters only need a single test path and only one cluster needs 4 paths. The total number of paths is 20. The paths are relatively short.

By the end of this step, we generated all the test paths for each cluster.

Table 13 Clusters test sequences.

ID	Test sequence	Length
— Main Page Test Sequences of Fig. 3		
1	[Main, **New Med**, Main, Exit]	4
2	[Main, **New Name**, Main, Exit]	4
3	[Main, **Modify Med**, Main, Exit]	4
	New Med Cluster Test Sequences of Fig. 4	
1	[**New medicines**, New Med, **New medicines**, New Med, **Dosage**, New Med, Cancel, Main Exit]	8
2	[**New medicines**, New Med, **When**, New Med, Cancel, Main Exit]	6
3	[**New medicines**, New Med, **Instructions**, New Med, Add New, Main Exit]	6
	New Name Cluster Test Sequence of Fig. 5	
1	[**New medicines**, New Name, **New medicines**, New Name, **New nameP**, New Name, Cancel, Main Exit]	8
2	[**New medicine**, New Name, **Dosage**, New Name, Cancel, Main Exit]	6
3	[**New medicines**, New Name, **When**, New Name, Cancel, Main Exit]	6
4	[**New medicines**, New Name, **Instructions**, New Name, Cancel, Main Exit]	6
	New NameP Cluster Test Sequences of Fig. 8	
1	[New NameP, Accept, New NameP, Cancel, New NameP]	5
	New Medicines Cluster Test Sequences of Fig. 9	
1	[New medicines, Accept, New medicines, Cancel, New medicines]	5
	Dosage Cluster Test Sequences of Fig. 10	
1	[Dosage, Accept, Dosage, Cancel, Dosage]	5
	When Cluster Test Sequences of Fig. 11	
1	[When, Accept, When, Cancel, When]	5
	Modify Med Cluster Test Sequences of Fig. 7	
1	[Modify Med, **Edit**, Modify Med]	3
2	[Modify Med, Cancel, Modify Med, Delete, Modify Med]	5

Table 13 Clusters test sequences.—cont'd

ID	Test sequence	Length
	Instructions Cluster Test Sequences of Fig. 12	
1	[Instructions, Accept, Instructions, Cancel, Instructions]	5
	Edit Cluster Test Sequences of Fig. 6	
1	[Edit, **Dosage**, Edit, Cancel, Edit]	5
2	[Edit, **When**, Edit, Cancel, Edit]	5
3	[Edit, **Instructions**, Edit, Cancel, Edit]	5

3.4.2 Path aggregation

The test sequences through FSM in HFSM are now aggregated into test sequences for the whole model. The same way FSMWeb does [17]. A number of aggregation criteria have been proposed: all-combinations, each choice and base choice coverage [62]. We apply all-combinations coverage. This is the most expensive aggregation coverage criteria. The process results in a set of aggregate paths. We call them *abstract tests*.

For example, the first test path in AFSM [Main, **New Med**, Main, Exit] in Table 13 can be aggregated as follows: The test path has one cluster node (New med). The cluster node should be replaced by the New Med cluster test paths. The results of this step are three paths:

1. Main, **New Medicines**, New Med, **New Medicines**, New Med, **Dosage**, New Med, Cancel, Main Exit, Main, Exit
2. Main, **New Medicines**, New Med, **When**, New Med, Cancel, Main Exit, Main, Exit
3. Main, **New Medicines**, New Med, **Instructions**, New Med, Add New, Main Exit, Main, Exit

Test 1 still has three cluster nodes: New Medicines, New Medicines and Dosage cluster nodes. Test 2 has New Medicines and When cluster nodes. Test 3 has New Medicines and Instructions cluster nodes. After we replace all the cluster nodes, we get the following test paths.

1. Main, New medicines, Accept, New medicines, Cancel, New medicines, New Med, New medicines, Accept, New medicines, Cancel, New medicines, New Med, Dosage, Accept, Dosage, Cancel, Dosage, New Med, Cancel, Main Exit, Main, Exit
2. Main, New medicines, Accept, New medicines, Cancel, New medicines, New Med, When, Accept, When, Cancel, When, New Med, Cancel, Main Exit, Main, Exit

3. Main, New medicines, Accept, New medicines, Cancel, New medicines, New Med, Instructions, Accept, Instructions, Cancel,
Instructions, New Med, Add New, Main Exit, Main, Exit

Table 14 shows the aggregated test paths of the Family Medicines list app as
sequences of nodes.

When we built the model, we added dummy nodes and transitions to
ensure single-entry-single exit cluster models. Table 14 shows them in
italics. These do not require any inputs as they are not really testing steps.
However, they increase the length of the test paths. The next step removes
these dummy nodes, and transitions and replaces each remaining node pair
(edge) with its corresponding input action constraint.

Algorithm 1 shows the procedure for test step reduction. The algorithm
has two inputs: the set of Aggregated Paths and the number of paths. The
output of the algorithm is the sequence of input constraints for each aggregated test path. The algorithm has two loops: the first loop processes all test
paths. The second loop visits each node pair (edge) of the test path and adds
the constraint on the edge to the sequence if there is one.

ALGORITHM 1 Test step reduction
Input: Set of Aggregated Paths, Number of Paths n.
Result: Sequence of Input constraints for Each Aggregated Test Path.

```
1: for i=1 to n do
2:    for j=1 to Length(path_i)-1 do
3:        if edge(node_j,node_{j+1}) has constraint then
4:            add to constraint sequence for path_i
5:        end if
6:    end for
7: end for
```

Table 15 shows the result of the test step reduction for the first aggregated test
path. Column one shows the id of the test path. Column two shows the
Transition Id for constraint sequence. Column three shows the Constraint
Sequence of the reduced test path. Each input action constraint is separated
by a horizontal line. The last column shows the length of the constraint sequence.
Note that test path length is reduced by more than half, from 23 to 11 steps.

One of the goals of this approach is to extend the FSMWeb to test mobile
applications while keeping the size and the complexity of test paths manageable. For this example, there are 11 test sequences, their length varying
between 4 and 11 with a total length of 87. Table 16 shows length before

Table 14 Aggregated test paths.

Id	Test path	Length
1	[Main, New medicines, *Accept*, New medicines, *Cancel*, *New medicines*, New Med, New medicines, *Accept*, New medicines, *Cancel*, *New medicines*, New Med, Dosage, *Accept*, Dosage, *Cancel*, Dosage, New Med, *Cancel*, *Main Exit*, Main, *Exit*]	23
2	[Main, New medicines, *Accept*, New medicines, *Cancel*, *New medicines*, New Med, When, *Accept*, When, *Cancel*, *When*, New Med, *Cancel*, *Main Exit*, Main, *Exit*]	17
3	[Main, New medicines, *Accept*, New medicines, *Cancel*, *New medicines*, New Med, Instructions, *Accept*, Instructions, *Cancel*, *Instructions*, New Med, *Add New*, *Main Exit*, Main, *Exit*]	17
4	[Main, New medicines, *Accept*, New medicines, *Cancel*, *New medicines*, New Name, New medicines, *Accept*, New medicines, *Cancel*, *New medicines*, New Name, New NameP, *Accept*, New NameP, *Cancel*, *New NameP*, New Name, *Cancel*, *Main Exit*, Main, *Exit*]	23
5	[Main, New medicines, *Accept*, New medicines, *Cancel*, *New medicines*, New Name, Dosage, *Accept*, Dosage, *Cancel*, *Dosage*, New Name, *Cancel*, *Main Exit*, Main, *Exit*]	17
6	[Main, New medicines, *Accept*, New medicines, *Cancel*, *New medicines*, New Name, When, *Accept*, When, *Cancel*, *When*, New Name, *Cancel*, *Main Exit*, Main, Exit]	17
7	[Main, New medicines, *Accept*, New medicines, *Cancel*, *New medicines*, New Name, Instructions, *Accept*, Instructions, *Cancel*, *Instructions*, New Name, *Add New*, *Main Exit*, Main, *Exit*]	17
8	[Main, Modify Med, *Cancel*, Modify Med, *Delete*, *Modify Med*, Main, *Exit*]	8
9	[Main, Modify Med, Edit, Dosage, *Accept*, Dosage, *Cancel*, *Dosage*, Edit, *Cancel*, *Edit*, *Modify Med*, Main, *Exit*]	14
10	[Main, Modify Med, Edit, When, *Accept*, When, *Cancel*, *When*, Edit, *Cancel*, *Edit*, *Modify Med*, Main, *Exit*]	14
11	[Main, Modify Med, Edit, Instructions, *Accept*, Instructions, *Cancel*, *Instructions*, Edit, *Cancel*, *Edit*, *Modify Med*, Main, *Exit*]	14
	Total Length	181

Column one shows the id of the test path. Column two shows the abstract test paths. Column three shows the length of the test paths in terms of nodes. The total length of the test paths is 181 nodes. The longest path consists of 23 nodes, but there are only two of those. The shortest has 8 nodes. Median length is 17 nodes.

Table 15 Test path after reduction step.

ID	Edge Id	Constraint	Length
1	A1	R(SelectN, buttonANM) S(SelectN, buttonANM) continue-use(SelectN)	11
	D1	R(parMed, buttonANMA) S(parMed, buttonANMA) Continue-use(parMed)	
	D2	O(parMed), R(buttonANMC) S(parMed, buttonANMC)	
	A1	R(buttonANM)	
	D1	R(parMed, buttonANMA) S(parMed, buttonANMA) Continue-use(parName)	
	D2	O(parMed), R(buttonANMC) S(parMed, buttonANMC)	
	H3	R(buttonAD)	
	E1	R(parDosage, buttonADA) S(parDosage, buttonADA) Continue-use(parDosage)	
	E2	O(parDosage), R(buttonADC) S(parDosage, buttonADC)	
	B6	R(buttonAM)	
	A4	R(buttonBack)	

Table 16 Length of before and after reduction step.

Test path ID	Before reduction step	After reduction step	Inputs	Actions
1	23	11	6	11
2	17	8	4	8
3	17	8	4	8
4	23	11	6	11
5	17	8	4	8
6	17	8	4	8
7	17	8	4	8
8	8	4	4	4
9	14	7	4	7
10	14	7	4	7
11	14	7	4	7
Total	181	87	48	87

and after the reduction step. Column 1 shows the id of the test path. Column 2 shows the length of the aggregated test path from Table 14 in terms of number of nodes. Column 3 shows the length of the test paths after the reduction as number of edges in order to compare the result with the other approaches in Section 4. Column 4 and 5 show the inputs and actions. Since, every step includes one action, the number of actions and test steps is the same. Input action constraints may or may not require inputs or multiple inputs before an action (like a button click) leads to a transition event. We will discuss Column 4 and Column 5 in more detail in Section 3.5. The last row of the table shows the total length of aggregated test paths as 181 nodes (170 edges) and the total length of the test sequence after the reduction step as 87 transitions. We reduce the transitions by 49%. After the reduction step, the longest test sequence consists of 11 transitions, but there are only 2 of those. The shortest has 4 transitions. The median length is 8 transitions. The reduction step helps to keep the size and complexity of the test sequences manageable.

3.5 Phase 3: Input selection

The final step of test generation is selecting inputs to replace the input constraints in the test sequence constructed in Section 3.4.2. The test values are selected by the test designer.

We do not require specific input domain coverage to keep the test designer free to make their own decision in this regard. For example, the test designer can generate input values by covering partitions or randomly selecting values from a list as long as input selection constraints are met. The test designer chooses values for related inputs. For example, the test designer chooses which medicine to match with the patient name. At the end of this step, we have a set of inputs for the execution phase. Table 17 shows the set of inputs for test path 8 of Table 14. Column one shows the constraint sequence, and column two shows the input values that meet the constraints. The last column explains each value. Test 8 has four inputs and five actions. The inputs are patient name (parname) and medicine name (parMed) which occurs twice in test 8. The input values are selected by the test designer. Both ad-hoc and coverage based value selection are possible. The

actions are select patient name (SelectN), select medicine (SelectM), Click delete button buttonD, click cancel button (buttonCE) and click back arrow to exit the mobile app (buttonBack).

Table 17 Test path 8 with input values.

Edge Id	Constraint	Value	Explanation
A3	O(SelectN), R(SelectM) S(SelectN, SelectM) Continue-use(SelectN, SelectM)	selectN = "Trev" selectM = "Asprin"	Random selection from name list Random selection from medicine list
I3	O(parName), R(parMed, buttonCE) S(parName, parMed, buttonCE) continue-use(parName, parMed)	parName = "Trev" parMed = "Asprin" buttonCE = click	Random selection from the database Random selection from the database Push Cancel Edit Button
I2	O(parName), R(parMed, buttonD) S(parName, parMed, buttonD) buttonD	parName = "Trev" parMed = "Asprin" buttonD = click	Random selection from the database Random selection from the database Push Delete Button
A4	R(buttonBack)	buttonBack = click	Push back arrow to exit the app

The total number of inputs of our example is 48, and the total number of actions is 87. There is a difference between the inputs and the actions because some of the test steps only require clicks (actions). The unique inputs are name (parname), medicine name (parMed), amount of dosage (parDosage), time of medicine (parWhen) and medicine instruction (parInstruction). There are five unique inputs, and 27 unique actions (Buttons or Combobox).

3.6 Phase 4: Execute and validate tests

Unlike FSMWeb which assumes that testers make their tests executable manually, for Mobile Apps many Automatic tools are available to run the test cases, as long as we know which inputs need to be used. For example, tests can be converted to Selenium or any of the other candidate execution environments. If we use Selenium, the Appium server executes the Selenium code and reports results, including pass or fail for each test. Appium is easy to set up and available open-source. Hence, we chose it

```
@Test
public void Test8() throws MalformedURLException {

    setUp();

    List <WebElement> compoboxName =
            wd.findElementsByAndroidUIAutomator("new
                UiSelector().className(\"android.widget.TextView\")");
    WebElement selectName = compoboxName.get(0);
    selectName.sendKeys("Trev");
    MobileElement EditMed = (MobileElement) wd
            .findElementByAndroidUIAutomator("new UiSelector().text(\"Asprin : 25 mg\")");
    EditMed.click();
    MobileElement Cancel = (MobileElement) wd
            .findElementByAndroidUIAutomator("new UiSelector().text(\"Cancel\")");
    Cancel.click();
    EditMed.click();
    MobileElement Delete = (MobileElement) wd
            .findElementByAndroidUIAutomator("new UiSelector().text(\"Delete\")");
    Delete.click();
    wd.pressKeyCode(AndroidKeyCode.BACK);

    // exit
    wd.quit();
}
```

Fig. 14 Test 8 in selenium.

to execute our test cases for the example.[g] Selenium [65] is an open source software testing framework for web and mobile applications. Selenium provides a test domain-specific language to write the tests such as Java, Python, C# and PHP. Selenium runs on Windows, Linux and MAC. Fig. 14 shows the Selenium code of Test 8 in Table 14. First, the test function calls setUP() in line 4. The setup() function connects Appium with the mobile device before executing each test case. Desired capabilities specify the set up for the Appium server as well as the test values. This includes the connection type, device name, operating system version (platformVersion), operating system (platformName), app package and app activity. They are sent to the Appium server by the Android Driver via a URL Connection.

Lines 6–9 in Fig. 14 find the combobox of the patient name then select the name. Selenium will look for the input types in Table 2 and save the reference to the input types using the findEelementByAndroidUIautomator function. Then, we can send the input values and the events using the reference of the object such as AddMed.click(). Lines 10—12 find the medicine name combobox and select the medicine to enter for the edit page. Lines 13—15 find the cancel element and perform the cancel action. Line 16 enters the edit page again because we reference the edit page. Lines 17—19 finds the delete button and performs the delete action, then goes back to the main page. Line 20 presses the back button on the Android device. Finally, line 23 disconnects the connection with the mobile app to start a new test case.

---

[g] For evaluations of App testing tools, see Choudhary et al. [63] and Lämsä [64].

The test cases were executed on a Samsung Edge 6 phone with Android version 7.0. Ten test cases passed and one failed. Test 9 failed because the app was unable to change the patient and medicine name. The execution time for the test suite is 10 min.

4. Comparing FSMApp and ESG

In Section 2, we identified one approach [23] that also performs Black–Box MBT for Mobile Apps. In this section, we compare FSMApp with this approach [23]. Later in Section 5, we perform a number of case study comparisons.

We used the Family Medicine App (Section 3.2) to compare both approaches. Section 4.1 introduces the ESG method [23]. Section 4.2 compares the results for both approaches.

4.1 Event sequence graph (ESG) method

de Cleva Farto et al. [23] used an Event Sequence Graph (ESG) to test mobile apps. Their approach consist of the following phases:
1. Create the Event Sequence Graph (ESG) test model. An Event Sequence Graph (ESG) is a directed graph which includes events (nodes)[h] and edges to connect the events. The nodes "start" and "end" of the graph represent the start and the end of node of the graph. The ESG does not include (multiple) inputs in the graph explicitly rather, they are modeled with decision tables nodes that are associated with nodes that are marked as double circles.
2. Generate paths and implement test cases from the ESG model. de Cleva Farto et al. [23] use TSD4WSC to generate the ESG model and complete Event Sequences (CESs). An CES is a linear sequence test path. The CESs are generated from ESG to cover all edges. Input and output values are determined using the decision table(s) mentioned above. Then, CESs are converted to Robotium. The input selection is ad-hoc from the decision table.
3. Execute implemented test cases with Robotium and collect data. The CESs are executed in the Android Virtual Device (AVD). Execution time is measured and faults are identified.

We apply this method to the Family Medicines List App. Fig. 15 shows part of the Event Sequence Graph of the Family Medicines List application.

---

[h] FSMApp refer to events as actions.

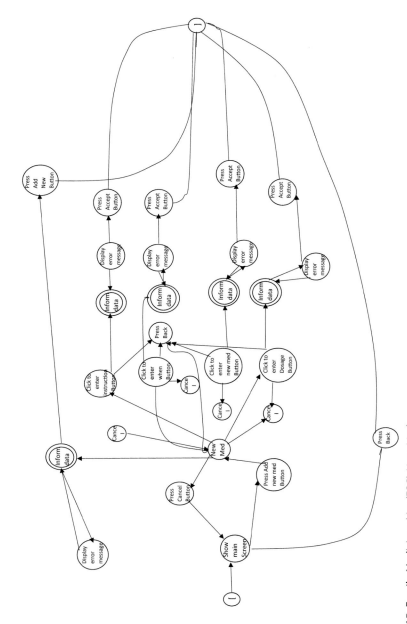

Fig. 15 Family Medicines List (ESG) New Med.

We divided the ESG into three figures since even for this small app, the graph is rather large. The double-circled nodes associated with providing a decision table(s). Decision tables (DT) describe type of input data required for test cases and any constraints for value selection [66]. Decision tables help to select inputs for the events of the ESG model. A decision table [67] is defined as $DT = \{C, E, R\}$ where
- C is a set of constraints with value true, false or do not care.
- E is a set of events.
- R is a set of rules for inputs that causes an event to occur.

This is similar to the input selection constraints in FSMApp. Table 18 shows only one DT of the family medicine list app because the limitation of pages. For example, Table 18 shows the decision table for new medicines. The constraints (C) are Medicine value (parMed), accept button (buttonANMA) and cancel button (buttonANMC). The Events are accept or cancel "add new medicine." The first rule (R1) means if perMed has a value and buttonANMA is clicked, then the medicine name is saved. R2 means regardless of when that perMed has a value and buttonANMC is clicked, then the medicine is cancelled. The decision table also covers error message events. If a value is selected that does not meet any of the constraint in the decision table or if a required field is not filled in, an error message is sent.

ESG generates single or multiple test paths that fulfil edge coverage. ESG generates 17 test paths ranging in length from 2 to 17 nodes. Table 19 lists the first five test paths.

Solving the input constraints in Table 19 results in the following inputs:
- $C1 = \{parMed = \text{"Aspirin"}; buttonANMA = Click\}$
- $C2 = \{buttonANMC = Click\}$

Table 18 New Medicines (ESG) decision table.

		Rules	
	New medicines	**R1**	**R2**
Constr.	parMed	T	DC
	buttonANMA	T	F
	buttonANMC	F	T
Events	Accept	C	
	Cancel		C

Table 19 Family Medicines List (ESG) Test Path.

Id	Test path	Length
1	[, Show_Main_Screen, Press_back,]	2
2	[, Show_Main_Screen, Press_Add_New_Med_Button, New_Med, Press_cancel_Button, Show_Main_Screen, Press_Add_New_Med_Button, New_Med, **Inform_Data**, Display_error_message, **Inform_Data**, Press_Add_New_Button,]	11
3	[, Show_Main_Screen, Press_Add_new_med_button, new_med, click_to_enter_instruction_button, cancel, new_med, click_to_enter_instruction_button, press_back, New_Med, Click_to_enter_Instruction_Button, **inform_data**, display_error_message, **inform_data**, press_accept_button,]	14
4	[, Show_Main_Screen, Press_Add_New_Med_Button, New_Med, Click_to_enter_when_button, cancel, New_Med, Click_to_enter_when_button, press_back, new_med, click_to_enter_when_button, **inform_data**, display_error_message, **inform_data**, press_Accept_Button,]	14
5	[, Show_Main_Screen, Press_Add_New_Med_Button, New_Med, Click_to_enter_med_button, cancel, new_med, click_to_enter_new_med_button, press_back, new_med, click_to_enter_new_med_button, **inform_data**, display_error_message, **inform_data**, press_accept_button,]	14

de Cleva Farto et al. [23] execute tests using the Robotium framework [68]. We use Appium so as to better compare the approaches. Table 21 shows the execution results. The execution time is 15 min. All test cases passed except two test paths failed (Test 5 and 8). ESG found one defect where the app cannot update the patient name and medicine name. The second, ESG test failed because when the test performs a press back button action to the previous state, the test setup failed, because the app exits instead of going to the previous page.

4.2 Comparison of results

Table 20 compares FSMApp and ESG with respect to model building, test generation, input selection, making tests executable and test execution. During model building, input constraints are represented on the edges in FSMApp whereas ESG represents them with double circle nodes with an

Table 20 Comparison of FSMApp and ESG.

Model building	FSMApp	ESG
Input Variables ($\overset{\rightarrow}{^{-}}I$)	part of edge annotation P($\overset{\rightarrow}{^{-}}I$, a) P($\overset{\rightarrow}{^{-}}I$, **a**) LAP	not explicit in node, annotated node with double circle.
Action (Event) (a)	edge	Separate decision table
App screen & node type	node type: cluster	node (annotated w/specific action value) implicit in node
Navigation		edge N/A
Cluster		
Test generation		
Cluster	local paths	N/A N/A
Coverage Criteria Test	node, edge, n–edge, prime path	from start node to end node
Paths	multiple paths via path	and reduced using a solution to
Coverage criteria	aggregation table all combinations, each choice, and base choice	the Chinese Postman Problem
Input Selection	resolve P along test path	ad-hoc from decision table, manual
Executable tests	manually; can use selenium/ Appium	Robotium
Test execution	manually; by tester exercising app can use selenium/Appium	By tester running Robotium

associated decision table. Actions are represented on the edges in FSMApp, but in ESG they are represented as a node. App Screen (Active) is a LAP in FSMApp, but a node in ESG. FSMApp and ESG navigate the model of the mobile application by edge traversal.

Clusters are only available in FSMApp. FSMApp generates the test paths by first generating paths for each cluster FSM, and then aggregating paths based on coverage criteria. ESG uses a tool to determine the test paths. Only FSMApp removes dummy nodes from test paths to reduce test sequences. Input generation

for test sequences in FSMApp is based on resolving input–action constraints. ESG uses ad–hoc selection from the decision table. In case of an incorrect value or a lack of filling a field an error message is sent.

FSMApp does not require to use any particular tools to make tests executable and execute them. Section 3.6 showed how to automate this step using Selenium/Appium. ESG runs tests using Robotium.

Table 21 Comparison of applying techniques in Family Medicines List App.

Step Id	Phase	Units of comparison	FSMApp	ESG
		Number of nodes	45	88
1	Generate Model	Number of edges	87	145
		Number of clusters	9	0
		Generation Time	14	29
		Number of test Sequences	11	17
2	Generate Test Sequence	Total test steps	87	219
		Generation time	31	38
Total time (1) + (2)			45	67
		#inputs	48	40
3	Input Selection	#actions	87	110
		Time to choose input	20	22
Total time (1) + (2) + (3)			65	98
		Test lines of code	547	635
4	Execute Test Cases	Execution time	11	15
		Number of failed tests	1	2
		Number of passed tests	10	15
		Number of defects	1	1
Total time			76	113

Table 21 compares the results of FSMApp and ESG for the Family Medicines App. Column one and two identify the four phases of test generation and execution. Phase 1 is model building. Phase 2 is test sequence generation, Phase 3 is input selection, and Phase 4 is execution and validation of tests. Column three shows units of comparison. For model generation, we compare model size in terms of nodes, edges, and clusters, as well as model generation time. For test sequence generation, we compare the size of the test sequence in terms of the number of sequences and the total number of test steps. We also compare the time it takes to generate them (in minutes). For input selection, we compare the number of inputs and actions, as well as input generation time. Finally, for test execution, we compare how much test code needed to be written, the number of tests that failed for passed and the number of defects found. We also compare execution time. Columns 4 and 5 show the results for FSMApp and ESG [23], respectively. We compare the results in Table 21 for each phase:

- Model generation: The FSMApp model is far smaller than the ESG model in terms of nodes and edges. Building the model for FSMApp takes much less time (12 vs 29 min). One reason for this is of course that the model for FSMApp is much smaller, but the cluster also required fewer repeated nodes and edges. FSMApp has half the number of nodes because the clusters reduce the number of nodes and edges instead of repeating them.

- Generation of test sequences: FSMApp generated 11 test sequences compared to 17 for ESG. FSMApp has significantly fewer total test steps (87 steps vs 219). ESG has far more steps because the approach results in many repeated actions and has many more test sequences. The FSMApp takes 31 min (compared to 88) because we generate the test sequences for each cluster, aggregate them, and then perform test step reduction.

- Input Selection: The total number of inputs and actions is comparable for ESG and FSMApp, with 135 vs 150 inputs and actions. Input generation times are also similar.

- Execution time: The test code required for FSMApp and ESG is roughly comparable (547 vs 635 LOC). ESG has a slightly higher execution time (15 compared to 11 min). They each find one defect. However, an additional ESG tests failed because when the test performs a press back button action to the previous state, the test setup failed, because the app exits instead of going to the previous page.

Next, we compare the overall time for testing (last row). ESG takes 113 min. This is much longer than for FSMApp which only takes 76 min. This is because model generation, time to choose inputs, and execution time for test cases required much less time than ESG. We can therefore conclude that FSMApp is more efficient.

5. FSMApp validation via case studies

5.1 Case study objectives

Now, we would like to compare and validate FSMApp and ESG for a larger number of mobile applications. The case studies cover mobile apps from different domains and with different sizes. We propose to investigate the applicability, scalability, efficiency and effectiveness of FSMApp for testing mobile applications. Furthermore, we want to know how FSMApp compares to ESG in these evaluation areas.

5.2 Case study research questions

The research questions derived from the case study objectives fall into two categories: those related to FSMApp and those related to comparison with ESG. Research Question RQ1-RQ4 deal with FSMApp, while RQ5 emphasizes comparison studies.

- RQ1: Applicability. Can we apply FSMApp to a variety of mobile apps in different application do- mains and of different sizes? Android Play [69] presents the top categories of mobile apps in the store: Photography, Family, Music & Audio, Entertainment, Shopping, Personalization, Social and Communication. AppBrain [70] presents slightly different top apps categories in the play store: Education, Business, Lifestyle, Entertainment, Music & Audio, Tools, Books & Reference, Personalization, Health & Fitness and Productivity. Some of the top categories overlap. The apps used in our case studies are taken from different categories to show the applicability of the FSMApp.
- RQ2: Scalability. How does FSMApp scale when models become larger? The case study includes multiple apps in the same category because we would like to test the scalability of the FSMApp with differ- ent sizes. The apps sizes on the disk range from 1.18 to 43.3 MB. However, the size of the app does not necessarily correlate with the model size needed for MBT. The case study compares the model size with respect to number of nodes, and number of edges.
- RQ3: Efficiency. How efficient is FSMApp? This is evaluated by steps in the test generation process, thus relates to efficiency of models (size), length of test paths and test suite, and test execution effort.
- RQ4: Effectiveness. How effective is FSMApp at finding defects for Mobile Apps of different sizes and in different domains? The case study executes the test cases and captures the number of the defects.
- RQ5: The following sub-questions compare FSMApp with other approaches using the same measurement as for RQ1–RQ4.
  RQ5.1: How does FSMApp compare to ESG in terms of applicability?
  RQ5.2: How does FSMApp compare to ESG in terms of scalability?
  RQ5.3: How does FSMApp compare to ESG in terms of efficiency?
  RQ5.4: How does FSMApp compare to ESG in terms of effectiveness?

5.3 Units of analysis

Table 22 shows the measurement units for every phase. Column one pre- sents the phase of FSMApp, and column two presents the measurement. These are the same metrics, we used in Section 4.2.

Table 22 Units of analyses.

Phase	Measurement
Generate Model	Size (#Nodes, #Edge, #Clusters), time to generate the model
Generate Test Sequences	Size (Number of test sequences, total test steps), time to generate the test sequences
Input selection	#inputs, #actions, Time to choose input
Execute Test Cases	Test line of code, Execution time, #fail, #success, #defects

5.4 Case study general descriptions & rationale

Table 23 summarizes the 10 Android mobile applications for the case study. The first column shows the name of the application. The second and the third columns show the number of user reviews and the rating of the application in Google Play. The fourth and fifth columns are related to the Android version that supports the App. The sixth and seventh columns show the size of the application code in terms of download size and size on disk. The eight column gives the type of the application. The last two columns provide the last update to the mobile application, and how many times the application was installed on mobile devices. The case studies cover seven domains: calendar, simple game, todo list, task management, music manager and File manager. We selected case studies that are Android open source apps with a review rating of 4 or over. The last three rows in Table 23 describe case studies that are available with source code from the Titanium development tool [79] or from student projects. The apps differ in size from small to larger to so we can validate the scalability of the FSMApp.

5.5 Case study results & discussion

5.5.1 RQ1: Applicability of FSMApp

We applied FSMApp to ten mobile applications from different categories. The apps fall into five catgories: Health & Fitness, Game, Music & Audio, Tools, and Productivity. We were successfully able to apply FSMApp to all ten mobile applications. We successfully applied FSMApp to all mobile apps in Table 23.

5.5.2 RQ2: Scalability of FSMApp

We apply FSMApp to mobile apps of varying size and application domain. We compare the FSMApp scalability with regards to four phases: Generate Model, Generate test sequence, input selection, and test execution and validation.

Table 23 Android Mobile Applications.

Apps	User Review	Rate/5	Android	Current version	Download Size	Code size (on Disk)	type	Last update	Installs	category
Family Medicines List [59]	Sample Code							02/25/2018		Health & Fitness
Memory Game Ap-plication [71]	Sample Code							02/25/2018		Game
Timber [72]	3044	4.3	4.1 and up	1.6	7.87 MB	5.14 MB	Music Player	12/12/2017	100–500k	Music & Audio
File Manager [73]	1303	4.1	4.2 and up	3.7	2.32 MB	1.18 MB	File Manager	11/19/2017	100–500k	Tools
ML Manager [74]	1345	4.6	4.1 and up	3.3	2.47 MB	3.48 MB	APK Extractor	01/23/2018	50–100k	Tools
Simple Calendar [75]	2634	4.3	4.1 and up	3.3.2	3.13 MB	4.32 MB	Calendar	02/22/2018	500–1M	Tools
Amaze File Manager [76]	9796	4.3	4.0 and up	3.2.1	4.59 MB	5.8 MB	File Manager	08/22/2017	500–1M	Tools
Todo List [58]	Sample Code							02/25/2018		Productivity
Minimal To Do [77]	725	4.5	4.1 and up	1.2	2.14 MB	12.3 MB	ToDo List	09/23/2015	10–50k	Productivity
MIRAKEL: Task Management [78]	281	4	4.0 and up	3	4.78 MB	43.3 MB	Task Management	07/29/2015	10–50k	Productivity

Fig. 16 shows the total number of edges vs the time to generate the model. Fig. 16 shows the time increases slowly but for the app whose model has 200 edges the time and reaches 60 min then go down to 30 min for a model with more edges. The time reaches 60 min because the behavior of the app is different and learning the app function for the first time needs more time. In general, since the tester' performance in the model build is measured, learning effects can occur.

Fig. 17 shows the total number of test sequences vs the time to generate the test sequences. Fig. 17 shows the time first decreases then it increases highly. There are two spikes. These apps have more clusters.

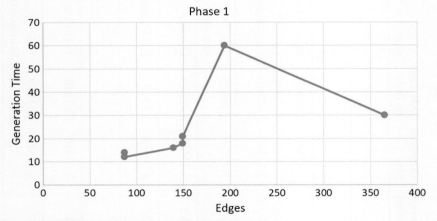

Fig. 16 Model generation time vs number of edges.

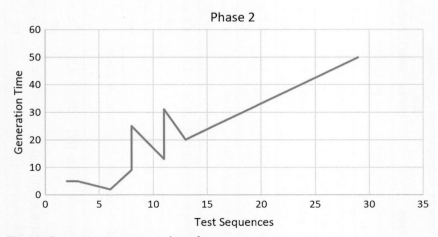

Fig. 17 Generation time vs number of test sequences.

Fig. 18 shows the total number of test inputs and actions vs the input selection time. The time increase linearly for less than 180 inputs and actions. It is increasing faster after 320 inputs. This depends on the number of components and types of inputs. The spike in the middle is related to clusters that have a lot of inputs which we do not consider them as nodes.

Fig. 19 shows the test LOC vs the execution/validation time. The execution/validation time increases rapidly with more than 1000 LOC. For lower LOC, the tests need less than 10 min to execute the test sequences.

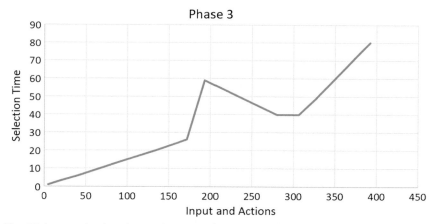

Fig. 18 Input selection time vs inputs and actions.

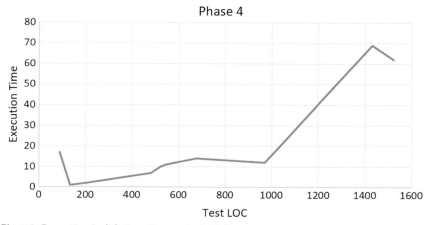

Fig. 19 Execution/validation time vs test LOC.

The time depends on the number of actions and how long it takes the mobile device to respond. It is important to know that even for the larger apps the execution of the tests is well below 2 h.

5.5.3 RQ3: Efficiency of FSMApp

We studied the efficiency of FSMApp. The efficiency is evaluated in the test generation process, length of test sequences, and test execution effort (time). Table 24 presents a summary for each app model size and time for generation model phase. Column 1 shows the name of the apps. Columns 2–4 show the number of nodes, edges, and clusters of the model, respectively. The last column shows the model generation time. Game Memory App [71] has the smallest model with six nodes and eight edges and no cluster. MIRAKAL app [78] has the largest model with 167 nodes, 465 nodes, and 12 clusters.

The model generation time of the 10 apps took between 2 to 60 min. The tool category has four apps: File Manager [73], ML Manager [74], Simple Calendar [75], and Amaze [76]. The four apps vary in the model size in term of nodes, edges and clusters. Simple Calendar has the smallest model size with 82 nodes, 149 edges, and 5 clusters, whereas File Manager, has the largest model with 188 nodes, 194 edges, and 18 clusters. The difference between the two models in term of size is around 30% for nodes and edges. We also include three apps from the Productivity category: Todo list [58], Minimal Todo [77] and task Management (MIRAKEL) [78]. Todo List has

Table 24 Model size summary.

App	Nodes	Edges	Clusters	Model time
Family Medicines	45	87	9	14
Memory Game	6	8	0	3
Timber	52	87	8	12
File Manager	118	194	18	60
ML Manager	87	149	7	18
Simple Calendar	82	149	5	21
Amaze	83	139	11	16
Todo list	12	18	1	2
Minimal	15	23	2	3
MIRAKEL	168	365	12	30

the smallest model with 12 nodes, 18 edges and one cluster whereas
MIRAKEL app has the largest model with 167 nodes, 365 edges and 12 clusters. The difference between the two models in term of size is 90%. FSMApp
can be applied to the same category app with many different sizes.

Table 25 shows the results of applying FSMApp to 10 mobile apps.
Column one indicates mobile apps name. Columns 2–4 show the number
of test sequences, number of steps (after reduction) and the time (in minutes)
for generating test sequence. Columns 5–7 show the number of inputs, number of actions, and time to choose inputs (in minutes) for the input selection
phase. Column 8 shows the execution time in minutes and Column 9 shows
the total time in minutes for all phases: model generation, test cases generation,
input selection, and test case execution. The last column shows the time per
step in seconds ((total time/number of steps) × 60). The time range of step for
all apps is between 27–56 s except the game app required 172 s per step. The
game app took a long time per step because the implementation of the game
has the images with a sources id. In this case, we use a loop to check all the
photos. The average time to execute each step is 42 s. The smallest model

Table 25 Summary of test generation and execution.

Apps	#test	#step	Time	#Input	#Actions	Choose time	Execution time	Total time	Step time
Family Medicines	11	87	31	48	87	20	11	76	52
Game Memory	6	8	2	0	5	1	17	23	172
Timber	8	120	9	3	190	59	7	87	44
File Manager	29	381	50	11	381	80	69	259	41
ML Manager	8	122	11	2	170	26	10	65	32
Simple Calendar	8	105	25	30	276	40	12	98	56
Amaze	11	188	13	14	266	40	14	83	27
Todo list	2	17	5	3	16	3	1	11	39
Minimal	3	20	5	6	35	6	2	16	48
MIRAKEL	13	277	20	19	308	49	62	161	35

required 48 s for each step. The largest model needed 41 s to execute the step, and the second largest model required 35 s. We can conclude that FSMApp is efficient for both large models and small models.

5.5.4 RQ4: Effectiveness of FSMApp

The case study executes the test cases and captures the number of defects. Table 26 shows the execution results for 10 apps. Column one shows the name of the mobile application. Columns 2 and 3 show the number tests that failed and passed, respectively. The last column shows the number of defects. The family Medicines app was unable to change the patient and medicine name whereas the app should allow the changes. All the apps do not show any defect except Family Medicines list app has one defect. This may be because we selected highly rated apps (see Section 7.3) and hence they are not likely to show many defects. The only except is the Family Medicines List app which was developed by student. Hence, with our selection of case studies, we were not able to show full effectiveness.

5.6 Compare FSMApp with ESG

5.6.1 RQ5.1: Applicability

FSMApp is can be applied in many domains and for varying sizes of mobile apps as show in Section 5.5.1. We applied ESG approach to all mobile apps in Table 23. Table 27 shows the result of ESG approach in Column 5.

Table 26 Defect summary.

App	#Failed	#Passed	#Defect
Family Medicines	1	11	1
Memory Game	0	1	0
Timber	0	8	0
File Manager	0	14	0
ML Manager	0	8	0
Simple Calendar	0	8	0
Amaze	0	11	0
Todo list	0	1	0
Minimal	0	3	0
MIRAKEL	0	13	0

Table 27 Comparison of applying techniques in Amaze File Manager.

Step Id	Phase	Unit of comparison	FSMApp	ESG
		Number of nodes	83	133
1	Generate Model	Number of edges	139	204
		Number of clusters	11	0
		Generation Time	16	26
		Number of test Sequences	11	54
2	Generate Test Sequence	Total test steps	188	384
		Generation time	13	25
Total time (1) + (2)			29	51
		#inputs	14	8
3	Input Selection	#actions	266	210
		Time to choose input	40	30
Total time (1) + (2) + (3)			69	81
		Test lines of code	676	1272
4	Execute Test Cases	Execution time	14	29
		Number of failed tests	0	0
		Number of passed tests	11	46
		Number of defects	0	0
Total time			83	110

Hence FSMApp and ESG have the same applicability. Table 27 reports the data of Amaze File Manager app only due to page limitations. They are organized the same as Table 21. They show the results for each phase: model generation, test sequence generation, input selection, and test case execution.

5.6.2 RQ5.2: Scalability

We compare the scalability of FSMApp and ESG with the model size in terms of edges, model generation time, number of test sequences, generation test sequences time, total of inputs and actions, time to choose input, test lines of code and execution time. Table 27 shows the comparison for the Amaze File Manager. The table is organized the same as Table 21.

Fig. 20 shows the total number of edges vs the time to generate the model for FSMApp and ESG. Fig. 20 shows the model generation time increases

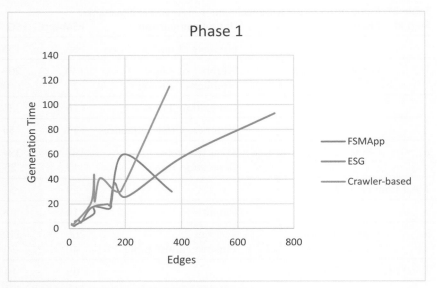

Fig. 20 Generation time vs number of edges.

more slowly for FSMApp than ESG resulting in 20 min for 150 edges. Then, the time reaches 60 min for FSMApp because the behavior of the app is different, and learning the app function for the first time takes longer. The ESG increases linearly beyond 200 edges. In general, since the tester's performance for model building is measured, learning effects can occur. If we exclude the data point with 60 min for a model with 200 edges for FSMApp, we have a linear line ranging between 20 to 30 min. Overall, Fig. 20 shows FSMApp is scalable compared to ESG in phase 1.

Fig. 21 shows the total number of test sequences vs the time to generate the test sequences. FSMApp has a maximum of 29 test sequences with a maximum time 50 min. We compare the efficiency and effectiveness of FSMApp and ESG approaches. The efficiency is evaluated for all phases of test generation and execution in Figs. 22 and 23. The number of sequences for ESG has a bigger increase than FSMApp because the number of the nodes is high. FSMApp applies clusters.

Fig. 18 shows the total number of test inputs and actions vs the input selection time. The time increases linearly for less than 180 inputs and actions. It is increasing faster after 320, inputs more so than ESG. We do not have any mobile apps with more than 500 inputs and actions. ESG has more than 500 inputs and actions to generate, and the time to do so increases linearly. ESG has a lot more inputs and actions because the number of test sequences is higher than for FSMApp.

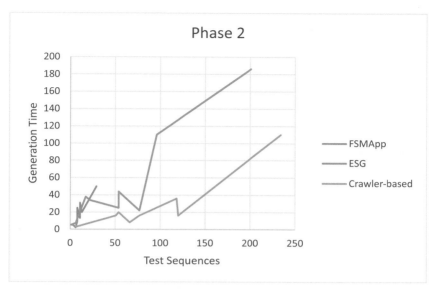

Fig. 21 Generation time vs number of test sequences.

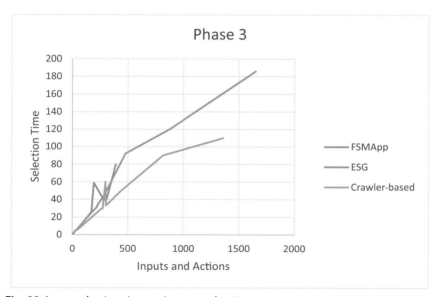

Fig. 22 Input selection time vs inputs and actions.

Fig. 19 shows the test LOC vs the execution/validation time. The execution/validation time increases rapidly with more than 1500 LOC for FSMApp and ESG. The largest app shows that FSMApp has the lowest test LOC because of the number of inputs and actions, and the number of test steps.

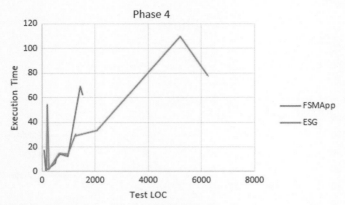

Fig. 23 Execution/validation time vs test LOC.

5.6.3 RQ5.3: Efficiency

In this section, we compare the efficiency of FSMApp with ESG. The efficiency is evaluated for all phases of test generation and execution.

The FSMApp model is almost half of the ESG model. Building the model for FSM takes much less time than ESG (3 vs 6 min), respectively. The reasons for this is the model for FSMApp is much smaller (6 nodes and 8 edges) than the model for ESG (12 nodes and 20 edges). The ESG model has twice the num- ber of nodes and more than double the edges compared to FSMApp. FSMApp generated 6 test sequences with eight steps, and the generation time is 2 min for the Memory game app. ESG generated one test sequence with 39 steps, and the generated time is 5 min. There is a big difference in the number of test steps between FSMApp and ESG. For FSMApp, the total time for the four phases is 23 min. The FSMApp takes less than half the time ESG.

We built the model for the Timber app with FSMApp in 12 min, whereas it takes almost 20 min to build the model with ESG. The difference is due to the clusters in FSMApp. The use of clusters reduces repeated nodes and edges. FSMApp has almost 40% less than ESG. FSMApp tests the Timber app with eight test sequences compared to 77 for ESG. ESG generates a large number of test sequences because there are many loops in the model. The FSMApp has fewer test steps compared to ESG (120 vs 592). The ESG approach has far more steps because the approach results in many repeated actions and has many more test sequences. The overall time for testing is 66 vs 125 min.

FSMApp reduces the model generation time and the model size by 70% compared to the ESG for the File Manager App. FSMApp uses 18 clusters to reduce the number of nodes and edges. The FSMApp model has 188 nodes, and 194 edges. The ESG model is much bigger (476 nodes and 732 edges). For the File Manager app, FSMApp generated few test sequences (29 vs 201). The FSMApp has fewer steps compared with ESG (381 vs 1451). The time to generate test cases is roughly half. The overall time for testing is 259 vs 489.

The ML Manager app shows different results than the other apps. The FSMApp model is a little bigger than ESG. Building the model for FSMApp takes more time than ESG (by 1 min). The reason for this result is that FSMApp has dummy nodes and edges between clusters. The ML Manager app can be tested by 8 test sequences vs 54 and 77 for ESG. FSMApp needs less than half the steps partly due to the test step reduction. FSMApp also needs fewer inputs and actions. The overall time is 65 vs 131 min.

Building a model of the Simple Calendar app for FSMApp and ESG takes roughly the same time, (21 vs 27 min). The model for FSMApp has 5 fewer nodes than ESG. Also, the model for FSMApp has fewer edge than ESG. The difference comes from dummy nodes and edges. Simple Calendar can be tested by 8 test sequences vs 22 for ESG. The number of test sequences for FSMApp is very small compared to ESG because ESG has many loops. FSMApp has half the steps of ESG because of test step reduction. There are also fewer inputs and actions. The overall test time is 88 vs 128 min.

The Amaze File Manager app results shows similar results to the other apps. The FSMApp model is more efficient and has fewer nodes and edges than the ESG. FSMApp generated fewer test sequences (11 vs 54). The FSMApp also has fewer test steps compared to ESG (188 vs 384). The time to generate test sequences is about half. The overall time for testing is 83 vs 110 min.

The Todo list app is the smallest app of the selected apps. The FSMApp model is smaller than the ESG model. FSMApp generated 2 test sequences with 17 steps, the test generation sequence time is 5 min. ESG generated 6 test sequences with 39 steps, and the rest sequence generation time is 6 min. There is a large difference in test steps between FSMApp and ESG. FSMApp's total test time of the four phases is 11 min vs 18 min.

For the minimal TODo app, FSMApp generated 3 test sequences with 20 test steps, with a test sequence generation time of 5 min. ESG generated 6 test sequences with 58 test steps, in 7 min. There is a large difference in test steps between FSMApp and ESG. The overall test time is 16 vs 25 min.

The FSMApp performs much better than ESG because the test sequences are short and the app is small.

For the MIRAKEL Test Management app, building the FSMApp model takes more time than ESG. The FSMApp model has more nodes and edges. The MIRAKEL app can be tested using FSMApp with 13 test sequences vs 96 for ESG. Also, the number of inputs and actions is much smaller. The overall test time is 151 vs 267 min. The difference stems from the fact that ESG needs more test steps and inputs.

5.6.4 RQ5.4: Effectiveness

The case study executes the test cases and captures the number of defects. FSMApp and ESG each found one defect. However, an additional ESG test failed because when the test performs a press back button action to the previous state, the test setup failed, because the app exits instead of going to the previous page. Also for ESG, one test failed in Simple Calendar, Timber, and MIRAKEL Task Management apps because of the same reason.

Overall, FSMApp compares favorably in effectiveness to the other two approaches. However, the high quality of the apps used in the case study makes conclusions related to effectiveness limited.

5.7 Threats to validity

We performed a number of case studies to evaluate the applicability, scalability, effectiveness, and efficiency of FSMApp to test Android mobile applications vs ESG [23].

We cannot yet generalize the results to other platforms. We only applied FSMApp to Android Applications and did not consider IOS and Windows. The second issue is the configuration of automation tools which test a mobile application for one Android device only. The third issue is the knowledge of the functionality of the tested mobile apps and how the functions are linked. Our choice of apps that have high rating make effectiveness conclusions limited. We should follow up with less robust apps.

We already compared the FSMApp results to the ESG approach. The cost of execution is calculated as a function of the number of steps in the test sequences. This may not be appropriate in all cases, since some steps, with more inputs to enter, and longer App execution time may affect results. However, [23] successfully applied the same approach. Further, the number of nodes can be affected by developer experience when generating test paths for the FSMApp and ESG.

Generalizability is limited as with any case study. We cannot guarantee that a future case performs and gives the same result for other Android applications, for example, advanced games, reservation applications, and significant medical applications. We showed with our case studies that the FSMApp could be applied to different categories of Android mobile application: a simple game, task management, file management, and music management.

Learning effects might bias the times needed to test each mobile app. One learning effect relates to the time it takes to understand how all functions in each mobile app work. If uncontrolled this could possibly lead to longer testing times for the first testing method applied to an app (generally this was FSMApp). To avoid this confounding factor, we studied all functions for each of the 10 mobile apps in detail to understand all components and the connections between the mobile screens, before applying the testing approaches. These learning effects were thus controlled by carefully analyzing how apps work before applying any of the testing methods.

Input constraints effectively partition the input space into input values that will cause a desired transition or event vs. those that do not. Often, there are multiple values that fulfill any given input constraint. We leave it up to the tester to select among those. This leaves the possibility (explored by Hamlet et al. [80]) that some selected inputs that meet the constraints will uncover a fault, but others may not. As this is the case for all methods studied, they all face the same issue. We tried to select similar values, when possible, to mitigate this problem.

6. Conclusions

This paper presents a black–box MBT technique to test mobile apps. FSMApp has four phases:

1. Generate Model: generate a hierarchical collection of FSMs models. This required developing a much larger number of input constraints than Andrews et al. [16] provide.
2. Generate Test Sequences: this involved generating paths through cluster FSMs and then aggregating them (like Andrews et al. [16]). Unlike Andrews et al. [16], we then reduced abstract tests to test sequences.
3. Input Selection: select the input values for each constraint in the test sequence.
4. Execute Test Cases: execute the test sequences with Appium.

We applied FSMApp to the Family Medicine list app. This paper, also compare FSMApp with two other approaches [23,36]. For the Family medicine list app, the FSMApp model is far smaller than the other two approaches in terms of model size. FSMApp also requires fewer test sequences and test steps. The total number of inputs and actions is comparable for ESG and FSMApp. ESG takes much longer. We can conclude that FSMApp is more efficient than ESG and equally effective.

This paper also used several case studies to investigate the applicability, scalability, efficiency, and effectiveness of FSMApp for testing mobile applications with ten mobile apps in different categories. Generally, FSMApp is more efficient, more scalable, while being equally effective.

In future work, we plan to extend FSMApp to selective regression testing. We also plan to study the effectiveness of FSMApp by applying it to case studies on pre-released apps.

References

[1] P.S. Kochhar, F. Thung, N. Nagappan, T. Zimmermann, D. Lo, Understanding the test automation culture of app developers, in: *2015 IEEE 8th International Conference on Software Testing, Verification and Validation (ICST)*, IEEE, 2015, pp. 1–10.
[2] "Number of mobile app downloads worldwide in 2017, 2018 and 2022 (in billions)," https://www.statista.com/statistics/271644/worldwide-free-and-paid-mobile-app-store-downloads/, accessed: 2018-11-02.
[3] M.D. Syer, M. Nagappan, B. Adams, A.E. Hassan, Studying the relationship between source code quality and mobile platform dependence, Softw. Qual. J. 23 (3) (2014) 485–508.
[4] A.I. Wasserman, Software engineering issues for mobile application development, in: *Proceedings of the FSE/SDP Workshop on Future of Software Engineering Research*, ACM, 2010, pp. 397–400.
[5] D. Zhang, B. Adipat, Challenges, methodologies, and issues in the usability testing of mobile applications, Int. J. Hum–Comput. Int. 18 (3) (2005) 293–308.
[6] H. Muccini, A. Di Francesco, P. Esposito, Software testing of mobile applications: challenges and future research directions, in: 2012 7th International Workshop on Automation of Software Test (AST), IEEE, 2012, pp. 29–35.
[7] J. Bo, L. Xiang, G. Xiaopeng, Mobiletest: a tool supporting automatic black box test for software on smart mobile devices, in: *Proceedings of the Second International Workshop on Automation of Software Test*, IEEE Computer Society, 2007, p. 8.
[8] D. Amalfitano, A.R. Fasolino, P. Tramontana, B.D. Ta, A.M. Memon, Mobiguitar: automated model-based testing of mobile apps, IEEE Softw. 32 (5) (2014) 53–59.
[9] A. Méndez-Porras, C. Quesada-López, M. Jenkins, Automated Testing of Mobile Applications: A Systematic Map and Review, in: Proc. 18th Ibero-Amer. Conf. Softw. Eng. Lima Peru, 2015, pp. 195–208.
[10] W. Yang, M.R. Prasad, T. Xie, A grey-box approach for automated GUI-model generation of mobile applications, in: Fundamental Approaches to Software Engineering, Springer, 2013, pp. 250–265.
[11] D. Amalfitano, A.R. Fasolino, P. Tramontana, S. De Carmine, A.M. Memon, Using GUI ripping for automated testing of android applications, in: *Proceedings of the 27th IEEE/ACM International Conference on Automated Software Engineering*, ACM, 2012, pp. 258–261.

[12] S. Anand, M. Naik, M.J. Harrold, H. Yang, Automated concolic testing of smartphone apps, in: *Proceedings of the ACM SIGSOFT 20th International Symposium on the Foundations of Software Engineering*, ACM, 2012, p. 59.

[13] C.S. Jensen, M.R. Prasad, A. Møller, Automated testing with targeted event sequence generation, in: *Proceedings of the 2013 International Symposium on Software Testing and Analysis*, ACM, 2013, pp. 67–77.

[14] C. Hu, I. Neamtiu, Automating GUI testing for android applications, in: *Proceedings of the 6th International Workshop on Automation of Software Test*, ACM, 2011, pp. 77–83.

[15] P. Wang, B. Liang, W. You, J. Li, W. Shi, Automatic android GUI traversal with high coverage, in: *2014 Fourth International Conference on Communication Systems and Network Technologies (CSNT)*, IEEE, 2014, pp. 1161–1166.

[16] A.A. Andrews, J. Offutt, R.T. Alexander, Testing web applications by modeling with FSMs, Softw. Syst. Model. (2005) 326–345.

[17] A.A. Andrews, S. Azghandi, O. Pilskalns, Regression testing of web applications using FSMWeb, in: Proceedings IASTED International Conference on Software Engineering and Applications, Nov. 2010.

[18] D. Kung, C.-H. Liu, P. Hsia, An object-oriented web test model for testing web applications, in: *Computer Software and Applications Conference. COMPSAC 2000. The 24th Annual International*, 2000, pp. 537–542.

[19] L. Ran, C. Dyerson, A. Andrews, R. Bryce, C. Mallery, Building test cases and oracles to automate the testing of web database applications, Inf. Softw. Technol. 51 (2009) 460–477.

[20] A.A. Andrews, J. Offutt, C. Dyreson, C.J. Mallery, K. Jerath, R. Alexander, Scalability issues with using FSMWeb to test web applications, Inf. Softw. Technol. 52 (1) (Jan. 2010) 52–66.

[21] L. Ran, C. Dyerson, A. Andrews, R. Bryce, C. Mallery, Building test cases and oracles to automate the testing of web database applications, Inf. Softw. Technol. 51 (2) (Feb. 2009) 460–477.

[22] A. Andrews, H. Do, Trade-off analysis for selective versus brute-force regression testing in FSMWeb, in: *2014 IEEE 15th International Symposium on High-Assurance Systems Engineering (HASE)*, Jan 2014, pp. 184–192.

[23] G. de Cleva Farto, A.T. Endo, Evaluating the model-based testing approach in the context of mobile applications, Electron. Notes Theor. Comput. Sci. 314 (2015) 3–21.

[24] C.D. Nguyen, A. Marchetto, P. Tonella, Combining model-based and combinatorial testing for effective test case generation, in: *Proceedings of the 2012 International Symposium on Software Testing and Analysis*, ACM, 2012, pp. 100–110.

[25] M. Utting, A. Pretschner, B. Legeard, A taxonomy of model-based testing approaches, Softw. Test. Verif. Reliab. 22 (5) (Aug. 2012) 297–312. [Online]. Available: https://doi.org/10.1002/stvr.456.

[26] A.C. Dias-Neto, G.H. Travassos, A picture from the model-based testing area: concepts, techniques, and challenges, Adv. Comput. 80 (2010) 45–120.

[27] H.-K. Miao, S.-B. Chen, H.-W. Zeng, Model-based testing for web applications, Jisuanji Xuebao(Chin. J. Comput.) 34 (6) (2011) 1012–1028.

[28] A. Marchetto, P. Tonella, F. Ricca, State-based testing of ajax web applications, in: In *2008 1st International Conference on Software Testing, Verification, and Validation*, IEEE, 2008, pp. 121–130.

[29] J. Ernits, R. Roo, J. Jacky, M. Veanes, Model-based testing of web applications using nmodel, in: Testing of Software and Communication Systems, Springer, 2009, pp. 211–216.

[30] H. Reza, K. Ogaard, A. Malge, A model based testing technique to test web applications using statecharts, in: *Information Technology: New Generations*, 2008. ITNG 2008. *Fifth International Conference on*, IEEE, 2008, pp. 183–188.

[31] F. Lebeau, B. Legeard, F. Peureux, A. Vernotte, Model-based vulnerability testing for web applications, in: Software Testing, Verification and Validation Workshops (ICSTW), 2013 IEEE Sixth International Conference on, IEEE, 2013, pp. 445–452.

[32] T. Pajunen, T. Takala, M. Katara, Model-based testing with a general purpose keyword-driven test automation framework, in: Software Testing, Verification and Validation Workshops (ICSTW), 2011 IEEE Fourth International Conference on, IEEE, 2011, pp. 242–251.

[33] B. García, J.C. Dueñas, Automated functional testing based on the navigation of web applications, EPTCS 61 (2011) 49–65, https://doi.org/10.4204/EPTCS.61.4.

[34] D.H. Nguyen, P. Strooper, J.G. Süß, Automated functionality testing through GUIs, in: Proceedings of the Thirty- Third Australasian Conferenc on Computer Science. vol. 102. Australian Computer Society, Inc, 2010, pp. 153–162.

[35] M. Sahinoglu, K. Incki, M.S. Aktas, Mobile application verification: a systematic mapping study, in: International Conference on Computational Science and Its Applications, Springer, 2015, pp. 147–163.

[36] D. Amalfitano, A.R. Fasolino, P. Tramontana, A GUI crawling-based technique for android mobile application testing, in: 2011 IEEE Fourth International Conference on Software Testing, Verification and Validation Workshops (ICSTW), IEEE, 2011, pp. 252–261.

[37] R.N. Zaeem, M.R. Prasad, S. Khurshid, Automated generation of oracles for testing user-interaction features of mobile apps, in: 2014 IEEE Seventh International Conference on Software Testing, Verification and Validation (ICST), IEEE, 2014, pp. 183–192.

[38] M.E. Delamaro, A.M.R. Vincenzi, J.C. Maldonado, A strategy to perform coverage testing of mobile applications, in: Proceedings of the 2006 International Workshop on Automation of Software Test, ACM, 2006, pp. 118–124.

[39] D. Amalfitano, A.R. Fasolino, P. Tramontana, N. Amatucci, Considering context events in event-based testing of mobile applications, in: 2013 IEEE Sixth International Conference on Software Testing, Verification and Validation Workshops (ICSTW), IEEE, 2013, pp. 126–133.

[40] S. Salva, S.R. Zafimiharisoa, Data vulnerability detection by security testing for android applications, in: Information Security for South Africa, 2013, IEEE, 2013, pp. 1–8.

[41] Y. Jing, G.-J. Ahn, H. Hu, Model-based conformance testing for android, in: Advances in Information and Computer Security, Springer, 2012, pp. 1–18.

[42] T. Azim, I. Neamtiu, Targeted and depth-first exploration for systematic testing of android apps, in: Acm Sigplan Notices, vol. 48, 10, ACM, 2013, pp. 641–660.

[43] R. Seiger, T. Schlegel, Test modeling for context-aware ubiquitous applications with feature petri nets, in: Proceedings of the Workshop on Model-Based Interactive Ubiquitous Systems (MODIQUITOUS), 2012.

[44] D. Franke, S. Kowalewski, C. Weise, N. Prakobkosol, Testing conformance of life cycle dependent properties of mobile applications, in: Software Testing, Verification and Validation (ICST), 2012 IEEE Fifth International Conference on, IEEE, 2012, pp. 241–250.

[45] S. Zein, N. Salleh, J. Grundy, A systematic mapping study of mobile application testing techniques, J. Syst. Softw. 117 (2016) 334–356.

[46] L. Lu, Y. Hong, Y. Huang, K. Su, Y. Yan, Activity page based functional test automation for android application, in: 2012 Third World Congress on Software Engineering (WCSE), IEEE, 2012, pp. 37–40.

[47] P. Costa, A.C. Paiva, M. Nabuco, Pattern based GUI testing for mobile applications, in: 2014 9th International Conference on the Quality of Information and Communications Technology (QUATIC), IEEE, 2014, pp. 66–74.

[48] C. Tao, J. Gao, Modeling mobile application test platform and environment: testing criteria and complexity analysis, in: Proceedings of the 2014 Workshop on Joining AcadeMiA and Industry Contributions to Test Automation and Model-Based Testing, ACM, 2014, pp. 28–33.

[49] A.K. Jha, S. Lee, W.J. Lee, Modeling and test case generation of inter-component communication in android, in: *2015 2nd ACM International Conference on Mobile Software Engineering and Systems (MOBILESoft)*, IEEE, 2015, pp. 113–116.

[50] Y.-M. Baek, D.-H. Bae, Automated model-based android GUI testing using multi-level GUI comparison criteria, in: *Proceedings of the 31st IEEE/ACM International Conference on Automated Software Engineering*, ACM, 2016, pp. 238–249.

[51] T. Takala, M. Katara, J. Harty, Experiences of system-level model-based GUI testing of an android application, in: 2011 Fourth IEEE International Conference on Software Testing, Verification and Validation, IEEE, 2011, pp. 377–386.

[52] Ui Automator. http://developer.android.com/tools/testing-support-library/index.html. accessed: 2016-01-26.

[53] Monkeyrunner. http://developer.android.com/tools/help/monkeyrunner_concepts.html. accessed: 2016-01-26.

[54] A.M. Memon, I. Banerjee, A. Nagarajan, GUI ripping: reverse engineering of graphical user interfaces for testing, in: Proceedings of the 10th Working Conference on Reverse Engineering, Nov. 2003.

[55] M. Nabuco, A.C. Paiva, R. Camacho, J.P. Faria, Inferring ui patterns with inductive logic programming, in: *2013 8th Iberian Conference on Information Systems and Technologies (CISTI)*, IEEE, 2013, pp. 1–5.

[56] A. Jaaskelainen, M. Katara, A. Kervinen, M. Maunumaa, T. Paakkonen, T. Takala, H. Virtanen, Automatic GUI test generation for smartphone applications-an evaluation, in: ICSE-Companion 2009. 31st International Conference on Software Engineering-Companion Volume, 2009, IEEE, 2009, pp. 112–122.

[57] A. Jääskeläinen, A. Kervinen, M. Katara, Creating a test model library for GUI testing of smartphone applications (short paper), in: *The Eighth International Conference on Quality Software*, 2008. QSIC'08, IEEE, 2008, pp. 276–282.

[58] Todo list. https://github.com/aaronksaunders/todolist.alloy. accessed: 2016-01-26.

[59] Family medicines list. http://basic384.rssing.com/chan-15081649/all_p2.html#item40. accessed: 2018-02-25.

[60] Anywhere software. https://www.b4x.com/b4a.html. accessed: 2018-10-18.

[61] Android. https://www.android.com/. accessed: 2018-10-17.

[62] P. Ammann, J. Offutt, Introduction to Software Testing, Cambridge University Press, 2016.

[63] S.R. Choudhary, A. Gorla, A. Orso, Automated test input generation for android: Are we there yet? in: *2015 30th IEEE/ACM International Conference on Automated Software Engineering Automated*, 2015.

[64] T. Lämsä, Comparison of GUI Testing Tools for Android Applications, University of Oulu, 2017.

[65] Selenium. http://www.seleniumhq.org/. accessed: 2018-01-20.

[66] G.J. Myers, C. Sandler, T. Badgett, The Art of Software Testing, John Wiley & Sons, 2011.

[67] M. Linschulte, On the role of test sequence length, model refinement, and test coverage for reliability, Universität Paderborn, Diss., Paderborn, 2013.

[68] Robotium. https://github.com/robotiumtech/robotium. accessed: 2016-01-26.

[69] Google play. https://play.google.com/store. accessed: 2019-01-14.

[70] Appbrain. https://www.appbrain.com/stats/android-market-app-categories. accessed: 2019-01-14.

[71] Memory game application. https://www.sourcecodester.com/android/8881/memory-game-application-android.html. accessed: 2018-02-25.

[72] Timber. https://play.google.com/store/apps/details?id=naman14.timber;https://github.com/naman14/Timber. accessed: 2018-02-25.

[73] File manager—storage, network, root manager. https://play.google.com/store/apps/details?id=dev.dworks.apps.anexplorer. https://github.com/1hakr/AnExplorer. accessed: 2018-02-25.

[74] Ml manager. https://play.google.com/store/apps/details?id=com.javiersantos.mlmanager; https://github.com/javiersantos/MLManager. accessed: 2018-02-25.

[75] Simple calender. https://play.google.com/store/apps/details?id=com.simplemobiletools. calendar;https://github.com/. SimpleMobileTools/Simple-Calendar, accessed: 2018-02-25.

[76] Amaze file manager. https://play.google.com/store/apps/details?id=com.amaze.file manager; https://github.com/. TeamAmaze/AmazeFileManager, accessed: 2018-02-25.

[77] Minimal todo. https://play.google.com/store/apps/details?id=com.avjindersinghsekhon. minimaltodo;https://github. com/avjinder/Minimal-Todo, accessed: 2018-02-25.

[78] Mirakel: Task management. https://play.google.com/store/apps/details?id=de.azapps. mirakelandroid&hl=en;https://. github.com/MirakelX/mirakel-android. accessed: 2018-02-25.

[79] Titanium mobile development environment. https://www.appcelerator.com/Titanium/. accessed: 2016-01-26.

[80] D. Hamlet, R. Taylor, Partition testing does not inspire confidence, in: Proceedings of the Second Workshop on Software Testing, Verification, and Analysis, IEEE, 1988, pp. 206–215.

About the authors

Dr. Ahmed Alhaddad is a graduate Ph.D. student at the Department of Computer Science/University of Denver, Denver, USA. His research is in regression testing.

Dr. Anneliese Andrews is Professor of Computer Science at the University of Denver. Before joining the University of Denver, she held the Huie Rogers Endowed Chair in Software Engineering at Washington State University. Dr. Andrews is the author of a text book and over 200 articles in the area of Software and Systems Engineering, particularly software testing, system quality and reliability. Dr. Andrews holds an MS and PhD from Duke University and a Dipl.-Inf. from the Technical University of Karlsruhe.

Zeinab Abdalla is a graduate Ph.D. student at the Department of Computer Science/ University of Denver, Denver, USA. His research is in regression testing.

CHAPTER TWO

Wheel tracks, rutting a new Oregon Trail: A survey of autonomous vehicle cybersecurity and survivability analysis research

Justin L. King, Elanor Jackson, Curtis Brinker, and Sahra Sedigh Sarvestani
Missouri University of Science and Technology, Rolla, MO, United States

Contents

Abstract

The rapid development of autonomous vehicles during the past decade has caused increasingly grave cybersecurity challenges to be associated with their use. Among these challenges are vulnerabilities involving existing vehicular technology, which have been subject to well-publicized exploits that bring into question the survivability of these vehicles under failure or attack. This chapter is a survey of the research landscape of autonomous vehicles, focusing on security and survivability; related attributes such as

1

Justin L. King et al.

performability are also considered. Research areas are visualized in a taxonomy and gaps are discussed throughout the paper. We conclude with recommendations and a discussion of future research opportunities.

1. Introduction

Many vehicles have the capability to augment human decision making, providing more information to the vehicle operator/driver than has been available at any point in automotive history. Technology enables drivers to safely change lanes and avoid collisions, but preventable accidents caused by driver fatigue, malicious intent, or carelessness remain an everyday occurrence. Poor decisions are made by drivers of all ages, but older adults have demonstrated greater difficulty dividing attention as compared to younger adults, leading to a direct correlation between their attention span and accident rates [1]. Factors such as chronic diseases and the use of medication can also impact cognition and are more prevalent among older drivers. As the population ages, the demand for fully *autonomous vehicles (AVs)* and AV-enabled technology will gradually increase. AVs could become more prevalent, at least initially by promoting the safety benefits and ability to give older adults as well as the disabled population a means of mobility. This opens up a new world of options previously unavailable to disabled persons, which could be used to provide reliable transportation to receive medical care, better employment opportunities, and leisure activities.

As older vehicles age and are replaced through attrition with newer technology, higher levels of autonomy are expected to become the norm. The vehicles on today's roadways span a wide range of automation, from limited features such as blind spot monitoring to lane centering and adaptive cruise control. A small fraction of vehicles exhibit higher levels of autonomy. This mixture of old and new technology compounds technical challenges as society moves toward acceptance and ultimately the desire to own and operate fully AVs.

The same technical challenges are present in military applications of AVs, where these vehicles can be instrumental in providing intelligence, surveillance, and reconnaissance and can ultimately reduce human casualties on the battlefield. Ubiquity of AVs could amount to a revolution in military affairs (RMA), requiring the assembly of a complex mix of tactical, organizational, doctrinal, and technological innovations in order to implement a new conceptual approach to warfare or to a specialized subbranch of warfare [2].

Although unmanned systems have been in existence since the mid–1900s, recent and rapid advancements in AV technology create an opportunity to catapult the next RMA.

In the United States, the Department of Defense (DoD) has made significant investments in unmanned systems. The most recent publicly available data is from fiscal year 2017, when the United States DoD budget request for development, procurement, and associated military construction (MILCON) of unmanned systems totaled more than $4B [3]. This figure is nearly a third of New Zealand's total military spending during the same period ($12.48B), nearly equal to Belgium's ($4.8B), and over twice that of the Netherlands' ($2.079B) [4].

With this technology comes the great responsibility of ensuring that unmanned systems operate securely, provide trustworthy data, and are survivable, i.e., can meet their mission requirements in both civilian and military applications. *Survivability* is a measure of a system's ability to continually deliver essential services during an undesirable event [5].

Autonomous vehicles are cyber-physical systems where sensors such as Global Positioning System (GPS) and LiDAR and a variety of actuators are interconnected through wired and wireless communication links. The Society of Automotive Engineers (SAE) categorizes vehicle autonomy on a scale that ranges from Level 0, no driver automation, such as a conventional vehicle, to Level 5, full autonomy, where the human is present to determine the operational readiness of the vehicle and merely to engage the automated driving system (ADS) [6]. This divergence of AV technology on and off the roadways and the complexity of communication paths available on vehicles present many challenges to survivability and availability. Fig. 1 highlights the sensing and communication technologies present in a typical autonomous vehicle. Full autonomy requires more advanced technologies to enable vehicle to vehicle (V2V) and vehicle to infrastructure (V2I) communications (discussed in greater detail in Section 2).

This chapter is a survey of literature related to the security and survivability of AVs, which we have classified based on the taxonomy depicted in Fig. 2. Published research on AV security is more abundant and primarily focused on identifying potential attacks and vulnerabilities and detecting intrusions. Research on survivability is far more limited. As of the date of publication, we were unable to identify any studies that examine an AV's ability to survive when facing large-scale cyber attacks or analyze the failure of an AV or multiple AVs to meet mission specifications. Research in cybersecurity and survivability is needed to ensure that AVs operate safely and predictably, so they can preserve life, increase public safety, and improve

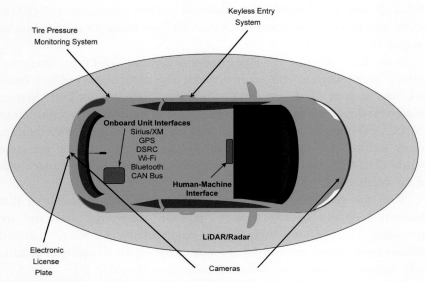

Fig. 1 Sensing and communication in an autonomous vehicle.

the mental wellbeing of drivers by reducing traffic congestion reduction. None of these goals are achievable without a survivable system.

2. Architecture

It is critical to first lay the foundation of the AV architecture before delving into more domain specific survivability and security topics. An in-depth discussion of an AV architecture would require at least a chapter, if not an entire book; however, that is not the focus of our research. We will attempt to be succinct as we work our way from the platform up to the *Intelligent Transportation System (ITS)*.

2.1 Platform

Fig. 1 depicts the architecture described in this subsection. An AV is a complex cyber-physical system whose safe operation requires complex software to process information from numerous hardware devices, from cameras to electronic license plates, all of which are connected to an electronic control unit (ECU). Increasingly powerful (or multiple) ECUs are being employed to enable functions such as automatic parallel parking [7]. Depending on the age and autonomy of the vehicle, 10 or more ECUs communicate with each other and the vehicle's Onboard Unit (OBU) over the Controller Area

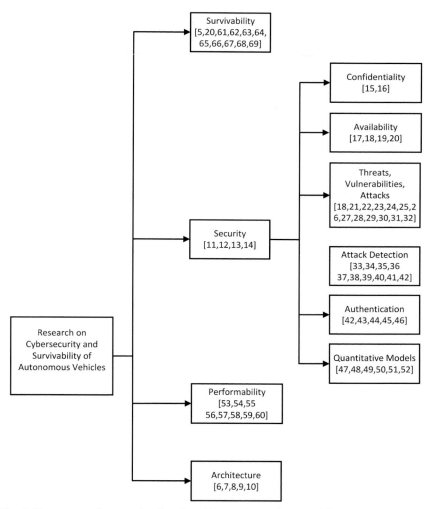

Fig. 2 Taxonomy of research related to AV security and survivability.

Network (CAN bus). Notable information is communicated with other vehicles and the infrastructure through a variety of message types sent over a wireless access in vehicular environments (WAVE) communication system, which is designed to meet the high-availability, low-latency communications requirements of vehicle safety functions such as precrash collision mitigation, intersection collision avoidance, and cooperative collision avoidance [8]. For critical applications such as forward collision warning (FCW), the OBU correlates data from sensors interconnected through the CAN bus in order to warn drivers of potential hazards from other vehicles ahead. Rear-end crashes are handled by the FCW application in three different scenarios: when the lead vehicle is stopped, moving at slower constant speed than the flow of traffic, or decelerating.

One critical message type is the SAE J2735 basic safety message (BSM). A BSM containing the vehicle's current position information (including latitude, longitude, speed, heading, and path history) is broadcasted every 100 ms over dedicated short range communications (DSRC) channel 172 [9]. DSRC allows for wireless communications among vehicles and between vehicles and the infrastructure. An alternative to DSRC for transmitting messages is cellular vehicle-to-everything (C-V2X), which communicates over cellular networks.

DSRC operates in the 5.9 GHz band, where the Federal Communications Commission (FCC) defines three levels of priority, as summarized in IEEE 1609.0 [10]:

1. The highest priority is *safety of life and property*.
2. The second highest priority is *public safety*.
3. The lowest level of priority applies to *private service providers*, including all commercial services.

State and local (nonfederal) government agencies fall under the "public safety" category, which includes groups such as first responders, highway maintenance, state and local departments of transportation, and public transit. Private ambulance services and volunteer fire departments may be included in this priority level through a memorandum of agreement.

Although "commercial services" are the lowest priority, they are the most critical in enabling the use of AVs. To facilitate development and adoption of AV technology, OBUs were designed with multiple interfaces, enabling multimodal connectivity for providing critical information to the operator and ITS. Intravehicle communication uses WiFi and Bluetooth to connect to a human–machine interface (HMI)—typically a tablet or a display unit built directly into the rear-view mirror. Intervehicle communications (outside of DSRC and C-V2X) involve GPS and SIRIUS/XM SATCOM.

2.2 Intelligent transportation system

An AV has two primary modes of communication with outside the vehicle: *vehicle-to-vehicle (V2V)* communication, which takes place directly between vehicles; and *vehicle-to-infrastructure (V2I)*, sometimes referred to as vehicle-to-roadside (V2R) communication. A host (ego) vehicle communicates with other vehicles (remote vehicle(s)) using DSRC. The sensors previously discussed aid in calculating the center of the lane that the vehicle is traveling on, the rate of curvature change, the lane width, and the relative position and velocity in relation to objects on and near the road; for a more in-depth treatment of this topic, see Ref. [9].

ITS infrastructure includes a DSRC device called a *roadside unit (RSU)*, which enables communications between vehicles and the larger ITS. The RSU communicates directly with a state's Transportation Management Center (TMC), to vehicles within range of the RSU (I2V), or to pedestrians (V2P) with a smart phone application. The TMC is a data center that provides enterprise-level services to the ITS and situational awareness to vehicles and pedestrians. TMC is the term designated in U.S. Department of Transportation (USDOT) Connected Vehicle (CV) Pilot Deployment Program resources. Fig. 3 depicts these potential communication channels between an AV and the ITS.

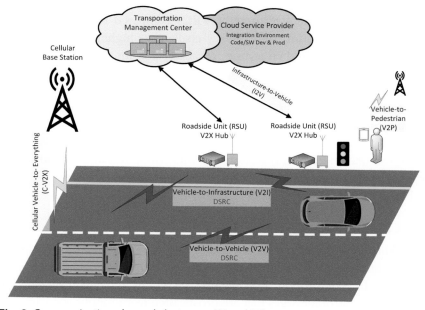

Fig. 3 Communication channels between AV and ITS.

3. Security

In their seminal publication on dependable and secure computing
[11], Avizienis et al. describe security as a composite of the attributes of confidentiality, integrity, and availability, requiring (1) availability for authorized actions only, (2) confidentiality, and (3) integrity. AVs present many
different security challenges. The threat actors against these cyber-physical
systems range from curious hobbyists, to lone wolf hackers, to nation states.

A wide range of studies on AV security, including several surveys, are
discussed throughout this section. One such survey by Ali et al. examines
security and privacy challenges of the ITS, rather than limiting the discussion
to AVs. The authors emphasize that an adversary can access an ITS user's
data without authorization, and that privacy mechanisms need to be
increased to protect users from compromise [12]. Schellekens broaches
the regulatory gap in automotive cybersecurity and the current state of laws
in the United States and European Union; he points out that current reg-
ulations do not satisfactorily address the AV security issue [13]. In Ref.
[14], the authors survey *vehicular edge computing (VEC)* and elaborate on
the specific architectural features of VECs compared to similar designs.
They also discuss security and privacy challenges, highlighting key unre-
solved issues such as authentication, reliability and integrity of data in the
cloud, and the employment of safety messages in order to provide a direct
trust relationship between vehicles. Increased awareness of AV threats and
vulnerabilities may accelerate regulatory developments that require vehicle
manufacturers and vendors of aftermarket devices to apply best practices for
security.

In the remainder of this section, we summarize research related to the
security of AVs. The grouping of studies is based on the taxonomy depicted
in Fig. 2.

3.1 Confidentiality

Confidentiality is the assurance that data is only revealed to authorized indi-
viduals. Encryption provides protection to data at rest and data in transit. In
Ref. [15], Li et al. propose a cooperative secret key agreement scheme for
encrypting control messages that are broadcasted from robotic vehicles in a
platoon. They state their intentions to extend this method to AVs equipped
with DSRC-enabled radios. Rathee et al. propose a novel approach to
address security by using blockchain to provide confidentiality, integrity,

and authentication for secure ride sharing, implemented on vehicular plat-forms and Internet of Things (IOT) devices [16].

3.2 Availability

Availability ensures timely and reliable access to and use of information; a loss of availability results in the disruption of access to or use of information or an information system [17]. In contrast to reliability, availability is able to account for system failure and repair. High availability is crucial for all critical infrastructure. The importance of availability is often overlooked, especially when systems remain operational for a long time. In AVs, where timely and reliable access to information from sensors and other sources is critical to safe operation of the vehicle, high availability is a critical requirement.

Studies on AV security, including surveys such as Refs. [18, 19], present only a short qualitative discussion of availability. We were unable to identify any quantitative AV-specific studies on availability and had to broaden the scope of our search to include wireless sensor networks (WSNs) to find quantitative analysis of availability. One such study, by Parvin et al., proposes an availability and survivability model using software rejuvenation for WSNs [20]. For a cluster of sensors with failure rate of λ_f, recovery (from failure state) rate of μ_f, and probability π_c of being in a compromised state, they define availability as:

$$1 - \left(\frac{\lambda_f}{\mu_f} \cdot \pi_c \right) \tag{1}$$

3.3 Threats, vulnerabilities, and attacks

Research on AV-specific threats, vulnerabilities, and attacks is abundant and appears to be growing at a rapid pace. Papers such as Ref. [21] discuss attacks against the ITS ecosystem. Similarly casting a wide net, Lu et al. present a comprehensive survey on the security of 5G vehicle-to-everything (V2X) services, with special focus on describing a plethora of attack vectors [18]. Many of the attacks discussed in this survey could also apply to V2V and V2I communications using DSRC. Sun et al. have authored a comprehensive survey on cybersecurity vulnerabilities of and risk classification for connected AVs. This study also covers defense strategies, standards, and open problems/topics for future research [22].

Moving one level deeper into attack vectors on the AV platform itself, Ren et al. give a comprehensive analysis of attacks that target an AV's sensors, authentication systems, and in-vehicle protocols; they also provide defensive strategies to counter these attacks [23]. In Ref. [24], their analysis is concerned with attacks against the in-vehicle network and ECUs. A study by Checkoway et al. provides a thorough analysis of vehicular attacks. Despite the wide scope of attacks discussed in this survey, little mention is made of GPS. The authors assert that GPS is an impractical access vector because the purpose-built hardware inherent to GPS transceivers will protect the signals from malicious activity [25]. The AV architecture has considerably advanced since the publication of this study (in 2011), with increased connectivity and a heavier reliance upon GPS; as such, greater caution must be used when ruling out any attack vector.

A narrower scope of attacks is explored by Palanca et al. in Ref. [26], which describes the implementation of a novel link layer denial-of-service (DoS) attack. Design weaknesses of the CAN protocol are exploited in this attack, which utilizes the standard diagnostic port. All CAN bus implementations are vulnerable to this exploit; however, the authors focused on the automotive domain, demonstrating their attack on an Alfa Romeo Giulietta. The exploit is possible due to the protocol mandate that all nodes that have lost arbitration shall in no way further interface with CAN traffic, leading to a denial of service. Amoozadeh et al. considered other attacks, including those that target cooperative adaptive cruise control in V2V communications, and found that the vehicles in their simulation were vulnerable to falsified beacon attacks [27]. Jafarnejad et al. focused on attacking the ECU of an experimental platform they built using a Renault Twizy 80, an open vehicle monitoring system (OVMS), and an Android application. The authors used the OVMS tool to access the vehicle's back seat computer over WiFi. The back seat computer was directly connected to the onboard diagnostics port (OBD) of the vehicle with an OBD-II cable. After brute force attack on the password, they were able to gain control of the safety-critical systems of the vehicle [28]. Oyler and Saiedian mapped and discussed threats against and risk mitigation strategies for the CAN bus, telematics, and related services [29]. A vehicle's telematics include the computer that has a built-in display as the operator interface. It typically provides radio, navigation, and hands-free cellular calling functionality; as well as entertainment such as game applications and DVD capability. Studies such as Ref. [29] are critical, given that telematics hardware is currently included as standard equipment on many entry-level vehicles. In this study, the authors categorize threats into the following categories:

- spoofing user identity
- tampering with data

- repudiation
- information disclosure
- denial of service
- elevation of privilege

In Ref. [30], Cheah et al. conduct CAN attacks and build on their prior work on systematic security evaluation to enumerate undesirable behaviors, enabling the assignment of severity ratings in a semiautomated manner. They describe challenges associated with evaluating the security of vehicles and discuss testing complications that arise from the complex architecture. The authors propose a four-stage testing methodology:

1. Model the threat manually using attack trees.
2. Conduct a penetration test using a semiautomated test execution tool.
3. Assign severity classifications in a semiautomated process.
4. Detail the security assurance case construction, conforming to SAE J3061. The authors illustrate their methodology by testing vehicle infotainment systems, where they were able to inject both OBD-II specific and raw CAN messages.

Sitawarin et al. examine two different types of attacks against sign recognition systems in Ref. [31]. The authors denote these attacks as deceiving autonomous caRs with toxic signs (DARTS) and demonstrate them in both virtual and real-world environment. The first attack portrayed is an out-of-distribution attack that exploits the fact that an image classifier will provide a classification for any input image. Knowing that the image classifier was trained to recognize traffic signs, the authors demonstrated that an adversary could mislead the classifier into mistaking a logo for a traffic sign. The second attack described is a lenticular printing attack, where the authors exploit the fact that a human operator and the AV's camera sensor observe the same sign from two different angles. They exploited this vulnerability by creating a special image from two different images and combining it with an array of magnifying lenses.

In Ref. [32], Shin et al. exploit vulnerabilities in LiDAR through a spoofing attack and a DoS attack. In the spoofing by relaying attack, the authors were able to induce fake dots closer than the spoofer's location—a task that had eluded past researchers. In traditional spoofing attacks, the target is fooled into believing that the originator or established communications are legitimate. In Ref. [32], the victim's LiDAR is tricked into believing that the detected object is closer than its actual real-world location. This is problematic, given LiDAR's importance in vehicle safety applications. The study also describes a novel saturation attack that results in the existing objects in the sensed output of the LiDAR. The authors thoroughly describe the process of the attack, tools used, and potential countermeasures to mitigate similar attacks.

3.4 Attack detection and prevention

Another cluster of research is centered upon detecting and preventing cyber attacks against AVs. When designing attack detection algorithms and devices, researchers must remain cognizant of how the AV will react once an anomaly is detected by the intrusion detection/prevention system (IDS/IPS). Key questions to consider are: (1) Who does the IDS notify? (2) How quickly does the notification need to occur? (3) Once the OBU is notified, can it employ other processes and ECUs in an attempt to recover from the attack? (4) Can the AV fail into a secure state? These issues can greatly impact the time to notification of the State's TMC, and more importantly, other drivers, when dealing with low data rate wireless links.

Much of the research in this area is moving toward computational intelligence techniques for attack detection. Tang et al. present a detailed survey on machine learning approaches to intrusion detection for vehicular networks and specifically identify machine learning methods for misuse detection and anomaly detection [33]. In a more traditional approach to attack detection and prevention, Harel et al. have identified constraints that prevent the use of tighter security measures for a vehicle's ECU [34]. This study also describes the design and implementation of in-memory validation and application whitelisting using BeagleBone Black hardware.

He et al. provide an overview of related work on AVs in order to build their case for the gap that they identify in current cybersecurity research [35]. Their research sets out to address the lack of a widely adopted AV cyberattack model; especially a model that can be easily adapted, defined, and classified consistently. The main goal of their research is to build a Unified Modeling Language (UML)-based framework for AVs to aid in the analysis of potential vulnerabilities. They also develop two different machine learning algorithms, which are used in their research to build models for anomalous behavior detection. For readers that are unfamiliar with the technology, the underlying communication architecture, and the most recent attacks on AVs, the paper does a good job at laying the groundwork, giving readers a better understanding of the breadth of potential attack vectors that an AV contends against. The authors described the scope of the problem in a tangible way by discussing the vast amount of data processed by a vehicle in 1 h (approximately 4000 GB), and by comparing the software lines of code in an AV (100 million) to a Boeing 757 Dreamliner (6.5 million). The authors adapt an extremely large data set

(KDD99) designed for traditional networked systems down to a more manageable level, tailored for AVs. The authors recognize that gaps still exist in their research, mainly because they note that decision tree and naive Bayes models using supervised learning cannot detect new attack types, and they state that their future work involves building clustering models to try to address this issue.

The biggest gap we see in this research is that the authors chose some attacks that do not apply to an AV, such as a mailbomb attack and warez client/server. This is partly due to fact that the KDD99 data set is over 20 years old, although it is one of the most widely used data sets for intrusion detection. They also left out a large attack surface in their research. The authors stated that "the paper only included communication-based attacks and not physical attacks." At first glance this may appear to be a small issue. Once the reader digs deeper into the way that the authors categorize physical parts, it includes sensors (such as LiDAR, RADAR, and cameras), global navigation satellite system (GNSS), OBD, CAN bus, and the AV power system. Despite this gap, they were able to find a way to make a meaningful impact at an extremely low cost. Cybersecurity testing on physical AVs is prohibitive from both cost and risk perspectives. It is also very challenging to find accurate data sets for testing AV cybersecurity attacks, given the constantly changing nature of the AV architecture.

Zhou et al. narrow the focus of their intrusion detection system, unlike the previous paper, by concentrating on anomalous CAN bus messages [36]. CAN messages are inherently insecure, like many of the other common protocols that were developed 40 years ago. CAN messages are broadcasted to the receiving ECU, with no way to validate the authenticity of the message. This opens the automotive network up to a variety of potential cyber attacks. The authors illustrate the security challenges with the CAN through a well-known example of a vehicle hack and further security architectural analysis that was presented by Valasek and Miller at respective Black Hat Conferences in 2013 and 2014 [37]. It is worth noting that Valasek and Miller published an additional paper in 2014, separate from the aforementioned Black Hat paper and presentation. In Ref. [38], they analyzed 14 different model year 2014 vehicles (varied by at least 2/3 criteria: manufacturer, make, and model), finding that 42% of the vehicles described in their survey had no separation between at least one cyber-physical ECU (one that would normally be segmented) and one that required external connections, expanding the vehicular attack surface. Zhou et al.'s research aimed to help solve this critical issue by developing a novel deep neural network (DNN) that detects anomalous CAN bus messages. They propose a DNN architecture with a triplet loss network that is designed in a pipeline fashion that optimizes the distance between the anchor example and the positive example; the goal is to make that distance smaller than the distance between the anchor example and the negative example. The authors describe a triplet loss function in deep learning as one that is typically fed positive and negative anchor samples, which improves object recognition. Their detection system uses CAN messages that are preprocessed offline for the anchor, which represents the annotated data. Real–time CAN messages are collected for the negative (abnormal data) and positive (normal data) pipes that are fed into the DNN. The three distinct data sequences from the same batch of messages are structurally consistent and share weights with each other. Each data set is then represented as three independent feature vectors. The data sets are composed of three parts: The training set, validating set, and test set. The authors used a tool that creates CAN messages, where they generated 200,000 packets for the data set, this breaks down into 150,000 normal messages and 50,000 abnormal. The last stage is the triplet loss function that is used to calculate the similarity and dissimilarity between random data sequences and the anchor.

The authors used the developed DNN and compared it to the performance of two other DNNs, DNN + support vector machine (SVM) and DNN + Softmax. Their performance evaluation showed that their model, the DNN + Triplet, outperformed the other methods in terms of detection accuracy, and the accuracy increases as the number of hidden layers increase. The one negative aspect that needs to be considered is the significant growth in detection times as the number of layers increase. At the top end of their analysis, using 16 hidden layers, it took 19 ms for detection at 98% accuracy. The authors did not specify the system requirements necessary to perform the evaluation. Given that DNNs typically require significant computational resources for CPU and GPU processing, it would have been worthwhile to understand if the test system is representative of a vehicle ECU or OBU architecture, which is most likely where the detection analysis would be performed, resident on the local AV.

Much like the previous paper, Park and Choi focus their paper on securing the CAN bus as well; however, they approach a different problem than the previous papers that were discussed [39]. They specifically look to address malware that impacts the Android operating system in vehicles.

They propose an architecture for an intrusion detection system using machine learning to detect this type of malicious behavior. The researchers utilized the Android Adware and General Malware data set for their machine learning-based IDS. The data set is described as containing traffic from 1900 applications downloaded from Google Play, in three categories: adware, general malware, and benign. The IDS contains four modules that include input, analysis, evaluation, and notification. The authors describe their method as an improved feature selection (IFS) method, combining the higher values derived from the correlation and information gain methods. The IDS module of the system has three phases:

- First, preprocessing is performed on the data set to select the most relevant features out of all the measuring features in the data set.
- Next, the data is modeled using a 10-fold cross validation correlation-based feature selection and an entropy-based information gain method; the modeled data is used to train the algorithm. Seventy-five percent of the data set is used for training, and 25% is used for testing and evaluation.
- Last is detection, where analysis is performed, learning, verifying, and evaluating message patterns using real in-vehicle network data.

In comparing their work to related studies, the authors employ the F_1 score, summarized below and described in detail in Ref. [40]. This metric is commonly used to evaluate the performance of binary classification algorithms and is defined as the harmonic mean of *precision* and *recall*. Precision refers to the fraction of positive classifications that are correct. Recall, also known as *sensitivity*, refers to the fraction of actual positives that are classified correctly. Eq. (2) summarizes commonly used performance measures for binary classification.

$$\text{Accuracy} = \frac{tp + tn}{tp + tn + fp + fn}$$

$$\text{Precision} = \frac{tp}{tp + fp}$$

$$\text{Recall} = \frac{tp}{tp + fn} \tag{2}$$

$$F_1 \text{ score} = 2 \times \frac{\text{Precision} \times \text{Recall}}{\text{Precision} + \text{Recall}}$$

The authors of Ref. [39] compare their proposed algorithm to six different machine learning algorithms: random forest (RF), decision tree (DT), *k*-nearest neighbor classifier (KC), gradient boosting classifier (GB), extra tree classifier (ET), and bagging classifier (BC). They discovered that the general method is not suitable for real-time detection because the elapsed time to detection was 3.750 s. The authors were able to calculate the F_1 score more quickly by doing the calculation when the model was in training. This significantly increased the speed of the algorithm to 0.049 s elapsed time, which outperformed all other algorithms. As far as performance, the random forest attained the best F_1 score, at 93.8% for multiclass classification and 92% for binary classification, but the authors' algorithm was not far behind, coming in at 92.9% combined. However, the speed of Park and Choi's algorithm is where it overshadowed the RF by beating it by more than a full second in both cases.

The authors used the full data set, which included 1500 top free apps from the Google Play Store (2015–2016). This casts a wide net, but it is worth experimenting on whether the accuracy and speed could have been improved had they narrowed the list down to applications that impact Android Auto. Another technique would be to create two separate algorithms, one that monitors Android Auto, and another to monitor traffic coming from USB, WiFi, and Bluetooth, which would cover smart phone connections.

Continuing with the CAN anomaly detection theme, Martinelli et al. analyzed real-world data sets that are freely available for research; the data set was generated using actual CAN traffic logged through the OBD-II port of a vehicle [41]. They focused on four categories of attacks: DoS, fuzzy attacks, spoofing the drive gear, and spoofing the RPM gauge. The feature vector was developed by aligning each vector to one byte of the CAN frame, resulting in eight feature vectors. Their evaluation of the messages went through three stages:

1. Compare the descriptive statistics of the normal and injected message populations.
2. Hypotheses testing to determine whether the feature vectors exhibit different distributions for attacks and normal message populations.
3. Classification analysis to assess whether the eight feature vectors discriminate between attacks and normal messages.

The four fuzzy algorithms for classification analysis are Fuzzy Rough Neural Network, Neural Network, Discernibility Classifier, and FURIA. The five metrics used to evaluate the results of the classification were false-positive rate, precision, recall, F-Measure, and receiver operating characteristic (ROC) area. They used 80% as testing and 20% of the data set as training, employing the full feature set. Their analysis determined that the Neural Network performed the best out of the four fuzzy algorithms. All the algorithms that the authors analyzed performed well when detecting normal and injected messages for the drive gear spoofing and RPM gauge spoofing attacks. The paper nor the source data set website provided enough details to determine exactly which parameters were spoofed for the gear and RPM test. This would help analyze why every one of the algorithms had 100% accuracy in detecting these types of attacks.

Kang and Kang utilized an unsupervised deep belief network (DBN) to train the parameters of their DNN, which they developed for intrusion detection of the CAN bus [42]. They provided examples of supervised artificial neural networks (ANNs) that were used in related research for CAN intrusion detection; however, they state that they are the first to propose an unsupervised algorithm for CAN IDS. For their DNN, they perform the training phase offline because of the considerable time impact. The system is trained by extracting a feature that represents a statistical behavior of the network. This feature is extracted directly from the bitstream of a CAN packet. The authors used the 64 bit positions in the DATA field of the CAN packet to generate the feature. The feature vector is then imputed into the input nodes of the structure, where an output with an activation function using a rectified linear unit (ReLU), and the linear combinations of the outputs are linked to the next hidden layers. The packets are then labeled as either normal or an attack packets. The output of the training phase is the weights that will be used as an initialization weight during the detection phase.

In the detection phase, the CAN packets used for testing are extracted in the same manner as was performed during the training phase. The output is then calculated with the weight from the training phase to determine whether the packet is a normal or an attack packet. Kang and Kang also used mode information to identify the test scenario, and value information to distinguish more granular parameters. Each vehicle's ECU has a unique identifier that distinguishes itself from the plethora of many other ECUs. This specific test identified three that they used, specific to the engine, body control, and display panel. They generated 200,000 packets in simulation with 70% of them being used for training and 30% for testing data. They varied the packets over time and used Gaussian noise in the value information to induce randomness in the test. The false-negative and false-positive rates were measured against an ANN and a SVM.

Their proposed method outperformed the other two methods by achieving a 97.8 total accuracy rate. Interestingly, when they compared the two deep learning methods against the number of layers, they discovered that the ANN suffered from the vanishing gradient problem, where the accuracy decreased as the number of layers increased. The accuracy rate of the proposed DNN IDS, however, improved as the number of layers increased. They also found that limiting the data to the specific modes and values was a way to improve upon the proposed method when compared to including all the bytes in the CAN packet DATA field. The authors found that the time required for the processing features stage was 8–9 μs, whereas the packet classification stage took 2–5 ms, adequate for real-time application.

The attack detection papers discussed in this section proposed novel solutions to address serious cybersecurity concerns, focusing in on the most critical components of AVs. The authors gave good insights into potential data set sources and proven computational intelligence methods. These can be used in future research to portray highly correlated attack vectors and to develop new techniques for detecting attacks against AVs.

3.5 Authentication

Authentication provides assurance for the authenticity of communications, whether referring to an entity/identity or data, which are the two general categories defined in the International Telecommunications Union's (ITU) Recommendation X.800 [43]:

- A *peer entity* authentication service is the use at the establishment of, or at times during, the data transfer phase of a connection to confirm the identities of one or more of the entities connected to one or more of the other entities.
- A *data origin* authentication service provides the corroboration of the source of a data unit; this does not provide protection against duplication or modification of data units.

The most common, simplest, yet weakest form of authentication is a username and password, although multifactor authentication is becoming more prevalent with the advent of smartphones and other technological advancements. A similar form of multifactor authentication utilizing a smartphone and/or biometrics could be applied to an AV, if the frequency of reauthentication is reasonable.

Mishra et al. present a chaotic map–based mutual authentication framework for vehicular cloud computing in order to address gaps in high computational resource utilization, communication efficiency, verification of security, and anonymity of the security scheme and user [44]. They compare other vehicular authentication schemes against theirs and utilize a simulation tool called Automated Validation of Internet Security Protocols and Applications (AVISPA) to their design against known attacks. They found that their scheme, when comparing to the four other existing schemes, outperformed all but one in computation costs; however, the communication cost was significantly lower than all others.

In Ref. [45], the authors propose an authentication scheme that aims to address a few key issues in vehicular communications:

1. Condensing an emergency vehicle's time from dispatch to arrival at the incident location
2. Reducing the data storage burden on RSUs
3. Protecting information from tampering.

Like Mishra et al., the authors of Ref. [45] focused on anonymity, but only for the regular, nonemergency vehicles. Their scheme first allows the RSUs and vehicles to initialize with a chameleon hash; a *certification authority (CA)* enforces the security parameters in this stage. Once the vehicles come within range of a RSU, they are able to register and authenticate with that RSU. If the time period is valid, the vehicle and RSU use a common secret key in order to establish communications. The CA is used to help maintain the anonymity of the communicating vehicle and RSU by obfuscating their true identities. For emergency vehicles, there is a similar authentication process between the emergency vehicle and the RSU. The emergency message is broadcasted to all vehicles in range, in order to yield the right of way to an emergency vehicle. This is critical in solving the second of three key issues identified in the study, when combined with a navigation system to detect the shortest path based on traffic conditions and the ability for RSUs to communicate with traffic signal controllers to maintain emergency vehicle progress toward the accident. Wu and Horng compared their method to four similar experiments using an EstiNet simulator. In evaluating time spent in each task, they outperformed the comparative research in establishing private communications, vehicle broadcast messaging, and in key updating. They only lagged in identity registration, which only applies when establishing communications with a RSU for the first time. Other authors, such as Xie et al., also used a simulation environment to evaluate the performance of the authentication scheme employed [46]. The scheme proposed in Ref. [46] achieves conditional privacy preservation with greater efficiency than previous comparisons; for safety critical messages, batch message verification and signature aggregation were introduced due to the broadcast nature of these messages.

3.6 Quantitative models

In Asplund et al.'s study on secure and dynamic group formations, the authors describe two incorrect group membership views, unsound and incomplete [47]. *Unsound* is where members who do not exist in the physical world are included in a group; incomplete is where vehicles exist in the area and should be part of the view, but are omitted. Using an abstract model of vehicle locations and their sensing capabilities, a method is formalized to detect violations and verify views for group membership. The paper also describes the architecture that was developed using a Satisfiability Modulo Theories (SMT) solver for online detection and verification to analyze group membership. Compared to other existing research, Asplund's is unique in the use of formal reasoning, supporting security, and fault tolerance. They identify three ITS applications that require coordination among vehicles: managed intersections, managed roundabouts, and in-vehicle platoons. The faults and attacks that they focus on in this study are communication failure, detection omission, location falsification, and Sybil attacks (vehicle invents multiple identities). The keystone of their work is verifying correctness of group views; this is done with a detector that checks for inconsistencies from other vehicles and data from internal sensors, which can also alert when an inconsistency is detected. The two basic properties of a view that they describe are soundness, where all vehicles in a view are located where the view claims they are located, and completeness, where all vehicles in an area are included in the view. The novelty of their method is the way nondetected alert data is fused with a model of the system and with the view to a verifier component that will assess whether the violation can or cannot be ruled out. This is a distinct change from related research, where the nondetected alerts were thrown out without the additional verification stage. The authors go on to formalize their method in great detail, and provide the pseudocode for view analysis, violation detection, and verification procedure. They provide a thorough evaluation of their algorithm, including use of a trace-based evaluation, in order to simulate a more realistic scenario. They concluded that in many cases, they were able to verify correct views equal to the ability to detect violations and hope that their work inspires a new paradigm in designing trustworthy coordination algorithms.

Although the research described in this paragraph is not specific to AVs, it addresses graphical security modeling techniques, which is a difficult task to undertake. The research described is also useful in modeling AV attacks and they also aid in calculating the probability of success and failure of the

specified attacks. Widel et al. provide an excellent survey of graphical security models, focusing on the application of formal methods to the interpre-
tation, semiautomated creation, and quantitative analysis of attack trees [48].
One such research paper that they describe in the survey is Kordy's work on
attack defense trees, which we will also discuss, given the magnitude of the
work on the ability to model attacks. Kordy et al.'s research provides a
computational framework for probabilistic evaluation of attack–defense
scenarios by developing an attack–defense tree using a graphical model
representation, combined with probabilistic information from a Bayesian
network model [49]. They keep the two models separate to allow for expert
input into each model. The other key difference between their work and
most other attack trees is that their model allows for dependencies, including
defense reactions to an attack. The output of both models is transformed
into an inference problem, where it can be solved by the fusion algorithm.
The authors developed a tool called ADTool to model the framework
described in their research. Gadyatskaya et al. took the existing ADTool
and completely revamped it to allow for ranking critical attack scenarios,
using attack trees with sequential AND (SAND) operator, and the addition
of scripting outside the GUI [50]. The SAND operator allows modeling of
attack steps that are executed sequentially, whereas before, the user could
only execute attack steps in parallel using the AND operator.

In Ref. [51], the authors propose a cyber risk classification model using a
Bayesian Network model, where values are derived from the *Common
Vulnerability Scoring System (CVSS)*. The CVSS score can be obtained from
the National Vulnerability Database (NVD). The NVD is funded by the
National Institute of Standards and Technology (NIST), and the most current is CVSS v3.1, although the authors listed the equation for the *Base Score*
using CVSS v2:

$$\text{Base Score} = (0.6 * \text{Impact}) - (0.4 * \text{Exploitability}) - (1.5 * f(\textit{Impact})) \quad (3)$$

$f(\textit{Impact}) = 0$ if the impact score is zero; otherwise $f(\textit{Impact}) = 1.176$.
The equation changed quite significantly in CVSS v3.1, where an
Impact Sub Score (ISS) is calculated before determining the subformula
values of impact and exploitability [52]:

$$ISS = 1 - (1 - \text{Confidentiality}) * (1 - \text{Integrity}) * (1 - \text{Availability}) \quad (4)$$

For the impact score, which feeds into the base score, NIST implemented
varying levels to account for whether the scope remains unchanged or if it

changed, as well as more granular calculation of exploitability. Sheehan used the proposed risk classification model and applied it to a GPS use case portraying two different attacks, jamming and spoofing. Their analysis calculated that GPS spoofing with no risk mitigation had the highest risk score (9.1); with cryptography, the risk lowered to 7.5. For the jamming attack, the risk score initially was an 8.6, which reduced to a 7.4 with cryptography. The model allows for expert input, as well as qualitative and quantitative data from 88,438 vulnerabilities from the NVD, which is used to refine the model and can also be used to predict new vulnerabilities.

4. Performability

Meyer, over 25 years ago, even while the intelligent transportation system was considered an emerging domain recognized the need to evaluate its performance. He described performance "in the typical use of the term in computer science and engineering as referring to how effectively (e.g., throughput, delay) or efficiently (e.g., resource utilization) a system delivers a specified service, presuming it is delivered correctly" [53]. Performability, combining performance and availability, allows for a varying degree of measurements, rather than a simple binary representation (operational or nonoperational state). In our group's previous work on survivability evaluation of cyber–physical smart grids, it was noted that survivability and performability have a similar objective—to characterize degraded operations. Evaluation of survivability and performability are equally necessary in AVs, especially while the opportunity to impact development exists. Given the finite resources and criticality of the time domain regarding AVs, some key metrics to consider are network and memory latency, hardware utilization, and CPU power consumption. Autonomous vehicles brought about computer architecture and organization challenges unlike any seen before in the history of computing. Key issues that challenge survivability and performability are the requirements for massive amounts of data storage and processing, artificial intelligence to assist and/or make decisions in real time, significant power consumption demands, and low bandwidth data rates combined with, at times, intermittent communications. All these issues are further exacerbated by vehicles traveling speeds, at times, in excess of 65 mph, which reduces the time to complete all of the aforementioned tasks. Additional research and testing is needed to address these concerns before AVs are more widely used; however, the window of opportunity to act is quickly closing, as demand and interest in AVs increase. This section surveys current AV performability-related literature regarding computer architecture, critical AV sensors, and communications such as the CAN bus and 802.11p. Some aspects of performability will be covered in the survivability section as well. We will also give recommendations and provide additional performance metrics based on similar critical systems.

Sensors, as discussed, in previous sections, play a pivotal role in observing the vehicle's environment where data coming off the various sensors must be aggregated, processed, and analyzed to enable the vehicle to make an intelligent decision. This requires significant processing resources in order to meet these demands, which in turn means increased power consumption. Performance metrics for AVs are at the initial stages of implementation, which need to be broad enough to be vendor and technology agnostic, but specific enough to measure. The National Highway Transportation Safety Administration (NHTSA) tried to implement such metrics by characterizing specific V2V communications requirements in Ref. [54]. The latest version shows that the document is out for comment, which is most likely attributed to the ongoing pilot initiatives in New York City, Tampa, and Wyoming. Some of the communication performance requirements that the NHTSA defined in the cited report, based on previous studies, are itemized below. These apply not only to DSRC, but to other competing technology as well, such as Connected Vehicle to Everything (C-V2X):

- DSRC data transmission rate for a BSM must be 6 Mbps.
- Initialization time within 2 s after the V2V device receives power.
- Packet Error Rate <10%.
- Stagger transmission times by randomly transmitting BSMs every 100 ms ±0–5 ms.

In their research, Behere and Torngren describe a functional reference architecture for autonomous driving. They provide an in-depth, logical architecture specific to autonomous driving. They make some interesting recommendations on future architectures based on previous experience on over 5 years worth of relevant projects. One that is worth mentioning, due to the significance commensurate with our topic, is that they recommend separating driving intelligence from the vehicular platform in order to lower cognitive complexity. This is not to say that it is off the vehicle completely, they reserve that for off-board guidance systems such as teleoperation, remote management, and fleet management. Other researchers architect an AI processor to handle computations required for neural networks. Their design integrates a Super-Threaded Core (STC) composed of 16,384 nano cores in a mesh-grid network, which they state performs at 32 TFLOPS. With the increasing concern of power management, their solution uses micropower gating for fine grain energy management, where the compiler generates STC commands based on the temperature profile [55].

Liu et al. also design an architecture to address the increasing computational and energy consumption requirements. A key metric in their article states that cameras used for sensing generate 1.8 GB of raw data per second, when combined. However, this did not include other sensors in their data rate analysis. The combination of multiple sensors compounds the issue of transmitting, receiving, and processing massive amounts of data on the platform. Their solution to address these demands in the computing platform layer included: A system on chip (SoC) architecture consisting of an I/O subsystem, a CPU, and shared memory through which the computing and I/O components communicate. The digital signal processor preprocesses the image stream to extract features; the GPU performs object recognition and other deep-learning tasks; the multi-core CPU is for planning, control, and interaction tasks; and the FPGA can be dynamically reconfigured and time-shared for data compression and uploading, object tracking, and traffic prediction [56].

In Ref. [57], the authors propose solving the sensor data issue by designing a hardware–software application, which processes incoming data from 3D sensors to detect pedestrians. They utilize a Point Cloud to convert the data before the application executes. Their design utilizes a machine learning system, which is hardware accelerated for preliminary data processing tasks, and the learning system is pure software. Their results showed that data processing tasks when running only in software without hardware acceleration took 103,460 ms compared to 344 ms with hardware acceleration.

Nou-Shene et al. sought to design an efficient, high-performance architecture to address real-time video stabilization consuming minimal power, and in a small form factor [58]. Even though their research is geared more toward AVs that are used in remote environments with rough terrain and sensitive to camera vibration, there remains applicability to AVs on city and state roads as well. Readers that are familiar with roadways in colder climates most likely have nightmare experiences when traveling, caused by the effects of treatments used to remedy snow and ice. The main culprit is potholes, and these craters wreak havoc on a vehicle and are sure to cause severe image stabilization issues on an AV's mounted camera system. This example is not limited to snowy locations though, potholes exist everywhere, as do gravel, brick, and even some cobblestone roads. These examples could cause the same impact. The authors describe the current technique for video stabilization as one that uses hardware accelerometers that feed information regarding the amount of vibration, which is an issue regarding accuracy and the speed at which the data is processed and validated.

Nou-Shene et al. propose an algorithm as well as a very large scale integration (VLSI) architecture to solve the stabilization problem. Each of the four stages of the algorithm are summarized below.

Local motion estimation module: First, static blocks are identified by their corner points. They used a block matching algorithm to calculate the displacement of the corner point compared to the previous frame. The displacement vector for each block is the estimated local motion vector of each corner point.

Global motion estimation module: Next, the global motion is computed from the local motion of the static blocks using a histogram-based approach.

Motion smoothing module: For this stage, a low-pass filter is used to remove unwanted disturbances, with the goal of retaining intentional motion.

Motion compensation module: This module obtains the smooth and stabilized video by offsetting the current frame by the accumulated motion vector.

The FPGA hardware that Nou-Shene et al. implemented for corner point detection and motion estimation are executed in parallel. Due to chip limitations, they were not able to store one complete frame on the chip. To address this issue, they utilized DDRAM to help avoid blank regions in the stabilization image. They setup a FPGA test "vehicle" in a lab environment and compared the design to two similar architectures, as closely as they could mimic the other researchers' work. Their analysis showed that the proposed method was more efficient than the other two designs with regard to area data global motion estimation and stabilization as well as power and time efficiency.

Li et al. investigated scheduling computing resources at the edge server for real-time applications in autonomous driving, considering vehicle mobility dynamics [59]. The authors define the age-of-results (AoR) as the traveled distance since the last data offloading before receiving the latest results delivery from an edge server, where the data being offloaded to the RSU is vehicle sensor data. Their sensor model assumes each processor handles a specific type of computing task. They describe two offloading techniques, synchronous and asynchronous. They describe synchronous offloading as when all the AVs connected to a RSU have the same sensing cycle and offload the sensor data to the RSU at the beginning of a computing cycle. Asynchronous is where the vehicles offload sensor data at arbitrary instants. The AVs' sensing cycle and the RSUs' computing cycle are of the same duration.

Li et al. use a Whittle index policy for the scheduling scheme and then compared it to a highest AoR policy (highest AoR schedule first) as well as a round robin policy (in circular order). For asynchronous offloading, they developed an algorithm that uses the AoR to guide scheduling. They fed the simulation a mobility table of varying speeds, with a computing cycle of 2 s, with 500 cycles ran to obtain the average AoR for all vehicles. They found that their proposed scheme in both synchronous and asynchronous scenarios performed the best among the tested policies. The researchers also used a vehicle mobility data set in a separate simulation running 3300 computing cycles while analyzing 126,376 sets of data. Their scheme performed as well as the SSA only scheme at 10–20 computing slots; however, the SSA only widens the performance gap as the number of slots increased. In terms of execution time, the running time was the highest of the four schemes; however, unlike the others, the author's scheme was able to adapt the time variant in real time. Their algorithm still executed the simulation when $K = 10$ at 0.12 ms and $K = 20$ at 0.16 ms, where K is the number of computing slots. This performance occurred simply using a general purpose system running an Intel i7-9750H CPU. The authors would likely see even more improvement with a more powerful system CPU and GPU due to the significant computing demands of the neural network that feeds the Whittle indices.

Yao et al. researched the safety-critical broadcast over the control channel of 802.11p [60]. The physical and data link layers of the WAVE standard are covered in IEEE 802.11p, which we briefly touched on in Section 2. The researchers modeled access categories using two Markov chains, where the broadcasted data was under saturated and nonsaturated conditions. They took advantage of the quality of service (QoS) options allowed in the MAC layer-enhanced distributed channel access. They defined four access categories with AC[0] being the highest priority messages reserved for emergency use messages. Given that they considered a bidirectional, one-lane highway, they developed a 1D VANET model, taking into consideration the maximum transmission range of 802.11p being up to 1 km.

A number of key assumptions were made in developing the models proposed in this paper:

1. The vehicles are exponentially distributed and satisfy the Poisson point process with density β (in vehicles per meter). The probability of having i vehicles in l meters of highway is given as:

$$P(i, l) = \frac{(\beta l)^i}{i!} e^{-\beta l} \tag{5}$$

2. Given a transmission range of R, interference range of L_{int}, and carrier sensing range of L_{cs} and assuming that $R \leq \lambda_{int} \leq \lambda_{cs}$, the average number of vehicles in the transmission range, the carrier sensing range, and the hidden terminal area, respectively, are given as:

$$\begin{cases} N_{tr} = 2\beta R \\ N_{cs} = 2\beta R \\ N_{ht} = 2\beta(R + L_{int} - L_{cs}) \end{cases} \tag{6}$$

Yao et al. assumed that the average packet size is the same for all access categories (ACs), with the packet arrival rates being different. They modeled each entity as a $M/G/1/K$ queue with a finite capacity, considering packets dropped, the buffer length is $K - 1$. In this notation, M denotes exponentially distributed interarrival times, G denotes a general service time distribution, 1 reflects a single server, and K reflects the total system capacity, including the buffer length of $K - 1$.

Unlike Li et al. [59], where they were highly focused on distance between vehicles, Yao et al. neglected the impact of vehicle mobility on the packet reception rate. They also assumed an ideal channel condition; therefore, they did not consider the packet error rate. The paper went into great detail regarding the models that they developed as well as their performance and survivability evaluation. They validated their work with the NS-2 simulator, using the transmission characteristics of 802.11p, and by providing highway scenario specifications. Under nonsaturated conditions, they used $\lambda_i = 5$ packets/s, and saturated conditions were $\lambda_i = 1000$ packets/s. They found that the collision probability was $< 10^{-9}$. They also determined that queue overflow was very small 10^{-4} to approximately 10^{-3}, but will rapidly increase under saturated conditions. Their analysis showed that 802.11p, given a typical highway scenario, satisfies delay constraints;

however, reliability requirements are not met. They note this as a research gap to be able to adjust parameters dynamically in the future, which would increase reliability.

5. Survivability

Imagine a scenario where emergency management personnel are deploying multiple AVs to transport citizens out of an evacuation zone during a natural disaster. The lead is informed through an emergency notification that there is a critical vulnerability on each of the AVs, OBU that needs to be patched immediately to prevent a known threat against the mission. In a rush to get the AVs operational before the next pickup time, the operator downloads the update from the Internet and then installs the update without following proper procedures to ensure the integrity of the update. Unbeknownst to the lead, the update introduced malware onto the OBU. The AVs are sent out for the mission, and then all of a sudden, communications are lost with the host/ego vehicle. The lead panics, not knowing how to handle the situation, but believes the AVs will come back online. After 30 min pass, the lead reports the chain of events to the emergency operations center. The following day, the lead finds out that the AVs were retrieved hundreds of miles from where they should have been located. The malware shut off communications between the AVs and the TMC as well as between the AVs and all RSUs, allowing a bad actor to take command and control to reroute them to a different destination, which were used to smuggle drugs through two states. The operator's decision impacted the survivability of the AVs, while redundant communications in the original system design could have prevented the failure. This notional scenario highlights the importance of cybersecurity and how it can impact survivability.

With survivability, the overall system operation can be measured at a degraded level, differing from other attributes such as dependability that are not as flexible. This gives survivability the means to quantify a varying degree of system operation. This flexibility makes survivability modeling ideal for critical infrastructure analysis as well as critical systems analysis. A critical system such as AVs operating on and connected to a critical infrastructure (the ITS) must also be able to deliver essential services during a fault or failure, including during and after cyber attacks. This was the original motivating factor for this survey. This section will specifically seek out AV survivability research. Survivability research is often scarce in many domains; therefore we supplemented with other pertinent survivability research where necessary.

Abdel-Rahim et al. analyzed the survivability of a large-scale ITS for Boise, Idaho using qualitative and quantitative methods [61]. They define survivability as "the capability of a system to fulfill its mission in a timely manner, even in the presence of component failures caused by intrusions, attacks, sabotage, accidents, or natural disasters." The qualitative method that they used was a modified version of Carnegie Mellon University's System Survivability Analysis (SSA). The major changes to the process were that they developed a stakeholder-by-responsibility matrix, they enumerated both physical and cyber threats, and they developed a threat-by-component matrix. The major finding from the qualitative analysis was that a loss of power supply at the Traffic Management Center (TMC) or at critical network intersections would have grave effects on the ability of the network to provide essential services. The authors did not address AVs specifically, but it is obvious that an AV's situational awareness would be impacted by each of the failures. Their findings highlight the necessity of redundancy, and effective disaster recovery planning. For their quantitative analysis, they used a multilayer graph-based approach; breaking the ITS into three layers: (1) the electric power grid layer, (2) the communications network layer, and (3) the physical roadway and control network layer. They defined the essentiality of each network component with its layer as:

$$e_\lambda(c) = p_\lambda - p_\lambda(c) \tag{7}$$

where $\epsilon_\lambda(c)$ is the essentiality of component c with respect to layer λ; p_λ is the performance of layer λ during normal operations, when no components have been damaged or disabled; $p_\lambda(c)$ is the performance of layer λ when component c is disabled.

The authors also assigned weights to components proportionally to their essentiality.

$$w_\lambda(c) = \frac{e_\lambda(c)}{t_\lambda} \tag{8}$$

where $e_\lambda(c)$ are the weights to component c with respect to layer λ; t_λ is the total of essentiality of all components under consideration.

The authors calculated the impact on network-wide travel times when there is a power failure at each of the 12 power substations that serve the ITS. Most of the substations, which serve multiple intersections, saw an over 80%

increase in network-wide total travel time, with only the two smaller substations seeing a 6% and 2% increase.

Other research used a different approach to address survivability. In Ref. [62], the authors focused on building a framework for autonomous unmanned ground vehicle (AUGV) survivability, where they meld biological needs and emotions to an autonomous system's survivability. In their article, they map Maslow's Hierarchy of Needs to an autonomous systems survivability needs. They define an AUGV's survivability needs as sustenance, safety, awareness, accomplishment, and cognition. The authors quantify the fulfillment of the needs of a system and then compute the controllability of the system, which assesses the ability to withstand and recover from any needs deficiency in finite time, with a finite sequence of actions. They also developed an optimization problem to maximize the likelihood that a system can operate autonomously over long periods of time. This work, although a little outside the norm, has relevance to AVs, especially where the authors defined the needs of autonomous systems.

Dharmaraja et al. identify a gap in recent literature regarding reliability and survivability of VANETs as a function of reliable hardware and channel availability [63]. The authors' modeling techniques include using reliability block diagrams, Markov chains, and Markov reward models to analyze V2V and vehicle to roadside communications. They modeled multiple reliability scenarios for both OBUs and RSUs. Their survivability analysis focused on modeling channel availability taking into account the control channel and service channels inherent in V2V communications; they also modeled channel failure, repair, and recovery. Other important survivability analysis that they conducted was to model the impact of channel unavailability and hardware failure of OBUs, the impact of channel failure and hardware failure of RSUs, the impact of hardware failure on the messages blocked and dropped, the connectivity of V2V and V2R communications as a function of reliable hardware and channel availability, as well as the expected number of messages lost during the recovery period of an OBU. They tested their analysis using MATLAB Simulink, over 100 time units, running the simulation 50 times with a 95% confidence interval. They found that over time, the reliability of the OBU and RSU decreased, impacting the number of messages lost. The number of messages lost occurred more for RSUs than it did for individual OBUs. This is due to the fact that RSUs communicate with multiple OBUs at a given time, thereby significantly increasing the number of failed messages.

In previous research, our group was concerned with survivability modeling, capturing the transient nature of critical infrastructures [64]. Similarly, Chang et al. [65] assess the survivability of a vulnerable critical system from the time that a severe vulnerability is announced through removal from the system. They use a homogeneous continuous-time Markov chain CTMC as well as a Stochastic Reward Net to automate the model. Their model would work well for an enterprise system that is connected to a high bandwidth medium, in order to scan and identify vulnerabilities. This article was not focused on AVs; however, the model would need to be modified for AV use to include rigorous testing of patches. Similar constraints would exist for any low bandwidth or air-gapped cyber-physical system. In Ref. [66], the authors were also concerned with capturing the transient nature of critical systems, with their research geared toward space systems. The metrics that they used to measure survivability were time-weighted average utility loss and threshold availability.

Mitchell and Chen were concerned with mobile cyber-physical systems (MCPS) surviving against malicious attacks, although they geared their paper toward battlefield or emergency response situations [67]. This distinction imposes unique power consumption constraints, requiring every piece of hardware and software to be highly scrutinized for their impact on power efficiency. They developed a mathematical model to assess the survivability of a MCPS equipped with a dynamic voting-based intrusion detection system. The attacks that they consider in their research are node capture and bad data injection. They measured survivability using mean time to failure (MTTF) of the MCPS as their metric for 128 sensor-carried nodes. They considered failure of the MCPS when energy is exhausted or when one third or greater of the total population of nodes have been compromised. The simulation batch size was 100 MTTF observations with the requirement to meet a 95% confidence level with 10% accuracy. Their analysis determined that with knowledge of per-node false alarm probabilities and per-node compromise rates, the system can dynamically select the best intrusion detection interval and the best number of detectors to balance energy consumption. This is an interesting undertaking to attempt to balance energy consumption and security. Further research would be warranted to analyze additional intrusion detection capability and security hardware/mechanisms such as a firewall and/or encryption impact energy consumption.

Parvin et al. took the availability formula that they developed from their stochastic model and expanded it to address survivability. Given that the wireless sensor network node is not survivable in the rejuvenation or failure state, they defined the survivability of each cluster of nodes as:

$$\text{Survivability} = \text{Availability} - (\pi_r + \pi_f) \qquad (9)$$

where π_r is the probability that the system is in rejuvenation state and π_f is the probability that the system is in failure state [20]. One concern for their basis of their threat model is that they assume that the base station cannot be compromised by an adversary, even though the base station is connected to the Internet via satellite communications. The authors do not explain the reasoning behind their assumption. However, in our case of AVs, which are critical systems, this would be a costly mistake as caution requires the assumption that compromise is imminent.

Knez et al. took a qualitative approach to cyber–physical survivability through the study that they conducted on behalf of the DoD Chief Information Officer/Cybersecurity (DoD CIO/CS). The study was focused on defining a systems security engineering (SSE) process, which included using existing artifacts such as the DoD System Survivability Key Performance Parameter (KPP)[68]. Three pillars are defined for cyber system survivability:

1. Prevent attack by reducing the likelihood of an attack.
2. Mitigate susceptibility by reducing the likelihood that an adversary will be able to complete an attack path.
3. Provide resilience by limiting mission harm due to cyber attack.

In Ref. [69], the focus is on resilience of autonomous systems, which they break down into resilience by design and resilience by reaction. Specific to our topic, the authors describe resilience of a ground and an aerial vehicle as a near horizon goal. They provide examples of advancements in the resilience of these areas, while mostly discussing open issues. These include resilience gaps such as time-varying characteristics like link quality on either platform, and the possibility of multiple coordinated and uncoordinated perturbations, with regard to their example of drones for medical deliveries where aerial and ground platforms must communicate (perturbations in this time-dependent case impact resilience). Another issue briefly described was the ever challenging problem of wireless medium signal propagation, which is compounded when dealing with subterranean environments. Subterranean environments posit their own challenges toward resilience beyond the scope of this chapter; however with a little background knowledge in the electro-magnetic spectrum, the reader can visualize potential issues with using extremely low or low frequencies to penetrate the soil and/or rock to extend communications. The most prevalent issue other than establishing reliable communications is the fact that the data rates are extremely low, which is why they must be reserved for special use cases.

6. Conclusion

The proliferation of autonomous vehicle technology is a growing area of concern for survivability in various sectors including public transportation, defense, and personal transportation. The research regarding these cyber-physical systems is increasing to meet the demand to provide solutions for complex transportation problems. These problems are compounded with poor design and antiquated protocols that were not designed for 21st Century vehicular technology, and especially not with cybersecurity built-in from inception. We surveyed cybersecurity and survivability attributes in order to map out the current landscape and to provide an opportunity to identify gaps where further research is needed. Survivability analysis remains a critical gap in AV research, and it is especially sparse when searching for experimentation while AVs are under cyber attack as well as faults leading to platform failure. Additional research is needed to model AVs under such challenged conditions, affording the opportunity to conduct further survivability analysis. Future work is necessary to provide a realistic means of diversifying AVs logical networks and protocols in order to increase survivability for the current generation of vehicles and to influence the near-term secure design of the next generation of AVs.

Acknowledgments

Funding from the Missouri University of Science and Technology Intelligent Systems Center, NSF DUE-1742523, and GAANN-P200A210121 is gratefully acknowledged.

References

[1] J. Yang, J.F. Coughlin, In-vehicle technology for self-driving cars: advantages and challenges for aging drivers, Int. J. of Automot. Technol. 15 (2) (2014) 333–340, https://doi.org/10.1007/s12239-014-0034-6.
[2] M. Knox, W. Murray, The Dynamics of Military Revolution 1300-2050, Cambridge University Press, 2001, ISBN: 0-521-80079-X.
[3] Unmanned Systems Integrated Roadmap 2017-2042, United States Department of Defense, 2017. https://news.usni.org/2018/08/30/pentagon-unmanned-systems-integrated-roadmap-2017-2042.
[4] CIA World Factbook, Central Intelligence Agency, 2021. https://www.cia.gov/the-world-factbook/.
[5] J.C. Knight, E.A. Strunk, K.J. Sullivan, Towards a rigorous definition of information system survivability, in: Proceedings DARPA Information Survivability Conference and Exposition, IEEE Comput. Soc, 2003, pp. 78–89, ISBN: 978-0-7695-1897-8, https://doi.org/10.1109/DISCEX.2003.1194874. http://ieeexplore.ieee.org/document/1194874/.

[6] SAE, Taxonomy and Definitions for Terms Related to Driving Automation Systems for On-Road Motor Vehicles, 2018, https://doi.org/10.4271/J3016_201806 (June).

[7] G. Rizzoni, Q. Ahmed, M. Arasu, P.S. Oruganti, Transformational technologies reshaping transportation—an academia perspective, in: Transformational Technologies Reshaping Transportation—An Academia Perspective, October, 2019, https://doi.org/10.4271/2019-01-2620. https://www.sae.org/content/2019-01-2620/. 2019–01-2620.

[8] SAE, V2X Communications Message Set Dictionary, (Generic), SAE International, Warrendale, PA, 2020, https://doi.org/10.4271/J2735_202007. https://go.exlibris.link/WzL6cw9n. (SAE Technical Standard).

[9] R. Miucic, Connected Vehicles: Intelligent Transportation Systems, Springer International Publishing, Cham, 2019, 978-3-319-94784-6 978-3-319-94785-3. https://doi.org/10.1007/978-3-319-94785-3. http://link.springer.com/10.1007/978-3-319-94785-3.

[10] IEEE, IEEE Std 1609.0-2019 (Revision of IEEE Std 1609.0-2013): IEEE Guide for Wireless Access in Vehicular Environments (WAVE) Architecture, IEEE Standard, 2019. (OCLC: 1104315709). https://ieeexplore.ieee.org/servlet/opac?punumber=8686443.

[11] A. Avizienis, J.-C. Laprie, B. Randell, C. Landwehr, Basic concepts and taxonomy of dependable and secure computing, IEEE Trans. Dependable Secure Comput. 1 (1) (2004) 11–33, https://doi.org/10.1109/TDSC.2004.2. http://ieeexplore.ieee.org/document/1335465/.

[12] Q. Ali, N. Ahmad, A. Malik, G. Ali, W. Rehman, Issues, challenges, and research opportunities in intelligent transport system for security and privacy, Appl. Sci. 8 (10) (2018) 1964, https://doi.org/10.3390/app8101964. http://www.mdpi.com/2076-3417/8/10/1964.

[13] M. Schellekens, Car hacking: navigating the regulatory landscape, Comput. Law Secur. Rev. 32 (2) (2016) 307–315, https://doi.org/10.1016/j.clsr.2015.12.019.

[14] R. Meneguette, R. De Grande, J. Ueyama, G.P.R. Filho, E. Madeira, Vehicular edge computing: architecture, resource management, security, and challenges, ACM Comput. Surv. 55 (1) (2023) 1–46, https://doi.org/10.1145/3485129.

[15] K. Li, W. Ni, Y. Emami, Y. Shen, R. Severino, D. Pereira, E. Tovar, Design and implementation of secret key agreement for platoon-based vehicular cyber-physical systems, ACM Trans. Cyber-Phys. Syst. 4 (2) (2020) 1–20, https://doi.org/10.1145/3365996.

[16] S. Rathee, A. Iqbal, K. Jaglan, A blockchain framework for securing connected and autonomous vehicles, Sensors 19 (14) (2019) 3165, https://doi.org/10.3390/s19143165. https://www.mdpi.com/1424-8220/19/14/3165.

[17] P. Bowen, J. Hash, M. Wilson, Information security handbook: a guide for managers, in: NIST SP 800-100, National Institute of Standards and Technology, 2006, p. 178.

[18] R. Lu, L. Zhang, J. Ni, Y. Fang, 5G vehicle-to-everything services: gearing up for security and privacy, Proc. IEEE 108 (2) (2020) 373–389, https://doi.org/10.1109/JPROC.2019.2948302. https://ieeexplore.ieee.org/document/8897696/.

[19] R.G. Engoulou, M. Bellaïche, S. Pierre, A. Quintero, VANET security surveys, Comput. Commun. 44 (2014) 1–13, https://doi.org/10.1016/j.comcom.2014.02.020.

[20] S. Parvin, D.S. Kim, S.M. Lee, J.S. Park, Achieving Availability and survivability in wireless sensor networks by software rejuvenation, in: Proceedings of the 4th International Workshop on Security, Privacy and Trust in Pervasive and Ubiquitous Computing, ACM, New York, NY, USA, 2008, pp. 13–18, ISBN: 978-1-60558-207-8, https://doi.org/10.1145/1387329.1387332.

[21] J. Harvey, S. Kumar, A survey of intelligent transportation systems security: challenges and solutions, in: 2020 IEEE 6th Intl Conference on Big Data Security on Cloud (BigDataSecurity), IEEE Intl Conference on High Performance and Smart Computing, (HPSC) and IEEE Intl Conference on Intelligent Data and Security

(IDS), May, IEEE, Baltimore, MD, USA, 2020, pp. 263–268, ISBN: 978-1-72816-873-9, https://doi.org/10.1109/BigDataSecurity-HPSC-IDS49724.2020.00055. https://ieeexplore.ieee.org/document/9123012/.

[22] X. Sun, F.R. Yu, P. Zhang, A survey on cyber-security of connected and autonomous vehicles (CAVs), IEEE Trans. Intell. Transp. Syst. (2021) 1–20, https://doi.org/10.1109/TITS.2021.3085297. https://ieeexplore.ieee.org/document/9447840/.

[23] K. Ren, Q. Wang, C. Wang, Z. Qin, X. Lin, The security of autonomous driving: threats, defenses, and future directions, Proc. IEEE 108 (2) (2020) 357–372, https://doi.org/10.1109/JPROC.2019.2948775. https://ieeexplore.ieee.org/document/8890622/.

[24] I. Studnia, V. Nicomette, E. Alata, Y. Deswarte, M. Kaaniche, Y. Laarouchi, Survey on security threats and protection mechanisms in embedded automotive networks, in: 2013 43rd Annual IEEE/IFIP Conference on Dependable Systems and Networks Workshop (DSN-W), June, IEEE, Budapest, Hungary, 2013, pp. 1–12, ISBN: 978-1-4799-0181-4, https://doi.org/10.1109/DSNW.2013.6615528. http://ieeexplore.ieee.org/document/6615528/.

[25] S. Checkoway, D. McCoy, B. Kantor, D. Anderson, H. Shacham, S. Savage, K. Koscher, A. Czeskis, F. Roesner, T. Kohno, Comprehensive experimental analyses of automotive attack surfaces, in: Proceedings of USENIX Technical Conference, 2011, pp. 77–92. 9781931971874; 1931971870.

[26] A. Palanca, E. Evenchick, F. Maggi, S. Zanero, A stealth, selective, link-layer denial-of-service attack against automotive networks, in: M. Polychronakis, M. Meier (Eds.), Detection of Intrusions and Malware, and Vulnerability Assessment, Lecture Notes in Computer Science, vol. 10327, Springer International Publishing, Cham, 2017, pp. 185–206, 978-3-319-60875-4 978-3-319-60876-1, https://doi.org/10.1007/978-3-319-60876-1_9. http://link.springer.com/.

[27] M. Amoozadeh, A. Raghuramu, C.-n. Chuah, D. Ghosal, H.M. Zhang, J. Rowe, K. Levitt, Security vulnerabilities of connected vehicle streams and their impact on cooperative driving, IEEE Commun. Mag. 53 (6) (2015) 126–132, https://doi.org/10.1109/MCOM.2015.7120028. http://ieeexplore.ieee.org/document/7120028/.

[28] S. Jafarnejad, L. Codeca, W. Bronzi, R. Frank, T. Engel, A car hacking experiment: when connectivity meets vulnerability, in: 2015 IEEE Globecom Workshops (GC Wkshps), December, IEEE, San Diego, CA, USA, 2015, pp. 1–6, ISBN: 978-1-4673-9526-7, https://doi.org/10.1109/GLOCOMW.2015.7413993. http://ieeexplore.ieee.org/document/7413993/.

[29] A. Oyler, H. Saiedian, Security in automotive telematics: a survey of threats and risk mitigation strategies to counter the existing and emerging attack vectors: security in telematics: existing and emerging attack vectors, Secur. Commun. Netw. 9 (17) (2016) 4330–4340, https://doi.org/10.1002/sec.1610.

[30] M. Cheah, S.A. Shaikh, J. Bryans, P. Wooderson, Building an automotive security assurance case using systematic security evaluations, Comput. Secur. 77 (2018) 360–379, https://doi.org/10.1016/j.cose.2018.04.008.

[31] C. Sitawarin, A.N. Bhagoji, A. Mosenia, M. Chiang, P. Mittal, DARTS: deceiving autonomous cars with toxic signs, arXiv:1802.06430 [cs] (2018). http://arxiv.org/abs/1802.06430.

[32] H. Shin, D. Kim, Y. Kwon, Y. Kim, Illusion and Dazzle: adversarial optical channel exploits against lidars for automotive applications, in: W. Fischer, N. Homma (Eds.), Cryptographic Hardware and Embedded Systems, Lecture Notes in Computer Science, vol. 10529, Springer International Publishing, Cham, 2017, pp. 445–467, ISBN: 978-3-319-66786-7, https://doi.org/10.1007/978-3-319-66787-4_22. http://link.springer.com.

[33] F. Tang, Y. Kawamoto, N. Kato, J. Liu, Future intelligent and secure vehicular network toward 6G: machine-learning approaches, Proc. IEEE 108 (2) (2020) 292–307, https://doi.org/10.1109/JPROC.2019.2954595. https://ieeexplore.ieee.org/document/8926369/.

[34] A. Kashani, G. Iyer, A. Kashani, G. Iyer, A. Harel, T. Ben David, A. Motonori, E. Masumi, Mitigating unknown cybersecurity threats in performance constrained electronic control units, in: WCX World Congress Experience, SAE International, 2018, https://doi.org/10.4271/2018-01-0016.

[35] Q. He, X. Meng, R. Qu, R. Xi, Machine learning-based detection for cyber security attacks on connected and autonomous vehicles, Mathematics 8 (8) (2020) 1311, https://doi.org/10.3390/math8081311. https://www.mdpi.com/2227-7390/8/8/1311.

[36] Zhou, Li, Shen, Anomaly detection of CAN bus messages using a deep neural network for autonomous vehicles, Appl. Sci. 9 (15) (2019) 3174, https://doi.org/10.3390/app9153174. https://www.mdpi.com/2076-3417/9/15/3174.

[37] C. Valasek, C. Miller, Adventures in Automotive Networks and Control Units, IOACTIVE, 2014. https://ioactive.com/resources/library/.

[38] C. Valasek, C. Miller, A Survey of Remote Automotive Attack Surfaces, IOACTIVE, 2014. https://ioactive.com/resources/library/.

[39] S. Park, J.-Y. Choi, Malware detection in self-driving vehicles using machine learning algorithms, J. Adv. Transp. 2020 (2020) 1–9, https://doi.org/10.1155/2020/3035741. https://www.hindawi.com/journals/jat/2020/3035741/.

[40] D.M.W. Powers, Evaluation: from precision, recall and F-measure to ROC, informedness, markedness and correlation, 2020. https://arxiv.org/abs/2010.16061.

[41] F. Martinelli, F. Mercaldo, V. Nardone, A. Santone, Car hacking identification through fuzzy logic algorithms, in: 2017 IEEE International Conference on Fuzzy Systems (FUZZ-IEEE), IEEE, 2017, pp. 1–7, ISBN: 978-1-5090-6034-4, https://doi.org/10.1109/FUZZ-IEEE.2017.8015464. http://ieeexplore.ieee.org/document/8015464/.

[42] M.-J. Kang, J.-W. Kang, Intrusion detection system using deep neural network for in-vehicle network security, PLOS ONE 11 (6) (2016) e0155781, https://doi.org/10.1371/journal.pone.0155781.

[43] International Telecommunication Union, X.800: Security Architecture for Open Systems Interconnection for CCITT Applications, International Telecommunications Union, 1991. https://www.itu.int/rec/T-REC-X.800-199103-I/en.

[44] D. Mishra, V. Kumar, D. Dharminder, S. Rana, SFVCC: chaotic map-based security framework for vehicular cloud computing, IET Intell. Transp. Syst. 14 (4) (2020) 241–249, https://doi.org/10.1049/iet-its.2019.0250. https://digital-library.theiet.org/content/journals/10.1049/iet-its.2019.0250.

[45] H.-T. Wu, G.-J. Horng, Establishing an intelligent transportation system with a network security mechanism in an internet of vehicle environment, IEEE Access 5 (2017) 19239–19247, https://doi.org/10.1109/ACCESS.2017.2752420. http://ieeexplore.ieee.org/document/8037972/.

[46] Y. Xie, F. Xu, D. Li, Y. Nie, Efficient message authentication scheme with conditional privacy-preserving and signature aggregation for vehicular cloud network, Wirel. Commun. Mob. Comput. 2018 (2018) 1–12, https://doi.org/10.1155/2018/1875489. https://www.hindawi.com/journals/wcmc/2018/1875489/.

[47] M. Asplund, Combining detection and verification for secure vehicular cooperation groups, ACM Trans. Cyber-Phys. Syst. 4 (1) (2020) 1–31, https://doi.org/10.1145/3322129.

[48] W. Wideł, M. Audinot, B. Fila, S. Pinchinat, Beyond 2014: formal methods for attack tree-based security modeling, ACM Comput. Surv. 52 (4) (2019) 1–36, https://doi.org/10.1145/3331524.

[49] B. Kordy, M. Pouly, P. Schweitzer, Probabilistic reasoning with graphical security models, Inform. Sci. 342 (2016) 111–131, https://doi.org/10.1016/j.ins.2016.01.010.

[50] O. Gadyatskaya, R. Jhawar, P. Kordy, K. Lounis, S. Mauw, R. Trujillo-Rasua, Attack trees for practical security assessment: ranking of attack scenarios with ADTool 2.0, in: G. Agha, B. Van Houdt (Eds.), Quantitative Evaluation of Systems, Lecture Notes

in Computer Science, vol. 9826, Springer International Publishing, Cham, 2016, pp. 159–162, ISBN: 978-3-319-43424-7, https://doi.org/10.1007/978-3-319-43425-4_10. http://link.springer.com.

[51] B. Sheehan, F. Murphy, M. Mullins, C. Ryan, Connected and autonomous vehicles: a cyber-risk classification framework, Transp. Res. Part A Policy Pract. 124 (2019) 523–536, https://doi.org/10.1016/j.tra.2018.06.033.

[52] FIRST, Common Vulnerability Scoring System v3.1: Specification Document, 2021. [online] Available: https://www.first.org/cvss/specification-document.

[53] J.F. Meyer, Performability evaluation: where it is and what lies ahead, in: Proceedings of 1995 IEEE International Computer Performance and Dependability Symposium, IEEE Comput. Soc. Press, 1995, pp. 334–343, ISBN: 978-0-8186-7059-6, https://doi.org/10.1109/IPDS.1995.395818. http://ieeexplore.ieee.org/document/395818/.

[54] NHTSA, Federal Motor Vehicle Safety Standards; V2V Communications, Department of Transportation, 2017. https://www.govinfo.gov/content/pkg/FR-2017-01-12/pdf/2016-31059.pdf.

[55] S. Behere, M. Törngren, A functional reference architecture for autonomous driving, Inf. Softw. Technol. 73 (2016) 136–150, https://doi.org/10.1016/j.infsof.2015.12.008.

[56] S. Liu, J. Tang, Z. Zhang, J.-L. Gaudiot, Computer architectures for autonomous driving, Computer 50 (8) (2017) 18–25, https://doi.org/10.1109/MC.2017.3001256. URL http://ieeexplore.ieee.org/document/7999133/.

[57] Q. Cabanes, B. Senouci, A. Ramdane-Cherif, A complete multi-CPU/FPGA-based design and prototyping methodology for autonomous vehicles: multiple object detection and recognition case study, in: 2019 International Conference on Artificial Intelligence in Information and Communication (ICAIIC), IEEE, 2019, pp. 158–163, ISBN: 978-1-5386-7822-0, https://doi.org/10.1109/ICAIIC.2019.8669047. https://ieeexplore.ieee.org/document/8669047/.

[58] T. Nou-Shene, V. Pudi, K. Sridharan, V. Thomas, J. Arthi, Very large-scale integration architecture for video stabilisation and implementation on a field programmable gate array-based autonomous vehicle, IET Comput. Vis. 9 (4) (2015) 559–569, https://doi.org/10.1049/iet-cvi.2014.0120.

[59] M. Li, J. Gao, L. Zhao, X. Shen, Adaptive computing scheduling for edge-assisted autonomous driving, IEEE Trans. Veh. Technol. 9 (2021) 5318–5331, https://doi.org/10.1109/TVT.2021.3062653. https://ieeexplore.ieee.org/document/9366426/.

[60] Y. Yao, L. Rao, X. Liu, Performance and reliability analysis of IEEE 802.11p safety communication in a highway environment, IEEE Trans. Veh. Technol. 62 (9) (2013) 4198–4212, https://doi.org/10.1109/TVT.2013.2284594. http://ieeexplore.ieee.org/document/6621048/.

[61] A. Abdel-Rahim, P. Oman, B. Johnson, L.-W. Tung, Survivability analysis of large-scale intelligent transportation system networks, Transp. Res. Rec. 2022 (1) (2007) 9–20, https://doi.org/10.3141/2022-02. http://journals.sagepub.com/doi/10.3141/2022-02.

[62] B.-K. Quek, J. Ibanez-Guzman, K.-W. Lim, A survivability framework for the development of autonomous unmanned systems, in: 2006 9th International Conference on Control, Automation, Robotics and Vision, IEEE, Singapore, 2006, pp. 1–6, ISBN: 978-1-4244-0341-7, https://doi.org/10.1109/ICARCV.2006.345336. http://ieeexplore.ieee.org/document/4150177/.

[63] S. Dharmaraja, R. Vinayak, K.S. Trivedi, Reliability and survivability of vehicular ad hoc networks: an analytical approach, Reliab. Eng. Syst. Saf. 153 (2016) 28–38, https://doi.org/10.1016/j.ress.2016.04.004.

[64] M. Woodard, K. Marashi, S. Sedigh Sarvestani, A.R. Hurson, Survivability evaluation and importance analysis for cyber-physical smart grids, Reliab. Eng. Syst. Saf. 210 (2021) 107479, https://doi.org/10.1016/j.ress.2021.107479.

[65] X. Chang, S. Lv, R.J. Rodriguez, K. Trivedi, Survivability model for security and dependability analysis of a vulnerable critical system, in: 2018 27th International Conference on Computer Communication and Networks (ICCCN), IEEE, 2018, pp. 1–6, ISBN: 978-1-5386-5156-8, https://doi.org/10.1109/ICCCN.2018.8487446. https://ieeexplore.ieee.org/document/8487446/.

[66] M.G. Richards, A.M. Ross, N.B. Shah, D.E. Hastings, Metrics for evaluating survivability in dynamic multi-attribute tradespace exploration, J. Spacecr. Rocket. 46 (5) (2009) 1049–1064, https://doi.org/10.2514/1.41433.

[67] R. Mitchell, I.-R. Chen, On survivability of mobile cyber physical systems with intrusion detection, Wirel. Pers. Commun. 68 (4) (2013) 1377–1391, https://doi.org/10.1007/s11277-012-0528-3.

[68] C. Knez, T. Llansó, D. Pearson, T. Schonfeld, K. Sotzen, Lessons Learned from Applying Cyber Risk Management and Survivability Concepts to a Space Mission, in: IEEE Aerospace Conference, March, 2016, pp. 1–8, https://doi.org/10.1109/AERO.2016.7500812.

[69] S. Bagchi, V. Aggarwal, S. Chaterji, F. Douglis, A.E. Gamal, J. Han, B.J. Henz, H. Hoffmann, S. Jana, M. Kulkarni, F.X. Lin, K. Marais, P. Mittal, S. Mou, X. Qiu, G. Scutari, Vision paper: grand challenges in resilience: autonomous system resilience through design and runtime measures, IEEE Open J. Comput. Soc. 1 (2020) 155–172, https://doi.org/10.1109/OJCS.2020.3006807. https://ieeexplore.ieee.org/document/9133332/.

About the authors

Justin King is a PhD Candidate in Computer Engineering at the Missouri University of Science and Technology. He holds a BS in Information Technology (Engineering Technology) from the University of Southern Mississippi, and an MS in Information Technology (Information Security) from Nova Southeastern University. He holds multiple cybersecurity industry certifications, including the Certified Information Systems Security Professional (CISSP) and Certified Ethical Hacker. Justin currently serves as a Supervisory General Engineer for the US Army Combat Capabilities Development Command (DEVCOM) Data and Analysis Center (DAC), after serving as the Chief Information Security Officer (CISO) for the Assistant Secretary of the Army for Acquisitions Logistics and Technologies (ASA(ALT)). He last served in the US Army Reserve 75th Innovation Command, prior to his retirement as a Lieutenant Colonel. His research interests include cybersecurity, autonomous vehicles and systems, cyber-physical systems, and machine learning techniques.

Curtis Brinker is working toward an MS degree in Computer Science at the Missouri University of Science and Technology. He earned a BS degree in Computer Science from the same institution in May 2022. His research interests include cyber-physical systems, autonomous systems and machine learning.

Elanor Jackson is a PhD student in Computer Engineering at the Missouri University of Science and Technology. She earned a BS in Computer Engineering from the same institution in May 2022. Her research interests include cybersecurity, autonomous vehicles, and modeling of complex systems.

Sahra Sedigh Sarvestani is an associate professor of Electrical and Computer Engineering at the Missouri University of Science and Technology. She holds a BSEE from the Sharif University of Technology, and an MSEE and PhD in electrical and computer engineering from Purdue University. Her research centers on analysis and modeling of dependability for complex systems, with focus on critical infrastructure and cyber-physical systems. Her research has been sponsored by the National Science Foundation, the US and Missouri Departments of Transportation, the Department of

Education, the National Security Agency, the European Commission, and private industry. She is a Fellow of the National Academy of Engineering's Frontiers of Engineering Education, a senior member of the IEEE, and a member of the IEEE Computer Society's Golden Core.

CHAPTER THREE

ClPyZ: A testbed for cloudlet federation

Muhammad Ziad Nayyer[a,b], Imran Raza[b], and Syed Asad Hussain[b]
[a]Department of Computer Science, GIFT University, Gujranwala, Pakistan
[b]Department of Computer Science, COMSATS University Islamabad (CUI), Lahore, Pakistan

Contents

Abstract

In the field of computing, the research conducted is largely of applied nature and for an applied research it is very vital to conduct experimentations and present results. These experimentations can be conducted using simulators, emulators, or real time testbeds. The data used to conduct these experiments can be synthetic or real. The researchers are often faced with challenges of finding a suitable instrument and environment to conduct and validate their research. This chapter presents a similar challenge of finding a suitable virtualization platform to mimic cloudlet federation and provides a viable solution by proposing a novel virtualization platform by the name of ClPyZ. The focus of this study is to highlight the fact that there is a lack of availability of simulators, emulators and virtualization platforms to conduct experimentations in this domain and a new virtualization platform is thus required. Cloudlet Computing is a variant of Mobile Edge Computing (MEC) that aims to provide the computational facility in closer proximity of the user to enhance Quality of Services (QoS) and Quality of Experience (QoE) for the resource constrained mobile devices. The concept of federation is used to pool resources among different cloudlets. ClPyZ does not only mimic a cloudlet federation but also offers resource sharing and load balancing at the cloudlet level to avoid remote request forwarding that results in improved performance. ClPyZ is an open source virtualization platform developed at Advanced Communication Networks Lab for research purposes. It can manage multiple cloudlets even if they belong to different clouds. A central broker manages all the affairs of the cloudlet federation such as cloudlet registration, finding an optimal cloudlet for task offloading and information exchange between member cloudlets. The required features for the inception of a federated cloudlet model include multi-cloud management, scalability, resource sharing, monitoring, Virtual Machine (VM) migration, and load balancing.

Abbreviations

ACLs	access control lists
APIs	application programming interfaces
CFRO	cloudlet federation for resource optimization
CLI	command line interface
CMA	cloud-based mobile augmentation
CMP	cloud management platform
CRM	cloudlet registration module
DaaS	database as a service
DSS	decision support system
EC	edge computing
EVP	enterprise virtualization platform
FC	fog computing
GUI	graphical user interface
IaaS	infrastructure as a service
IMM	information management module
IoT	internet of things
LAN	local area network
LTS	long term support
M2M	machine-2-machine
MCC	mobile cloud computing
MEC	mobile edge computing
OS	operating system
OVA	open virtualization appliance
OVB	oracle virtual box
PaaS	platform as a service
QoE	quality of experience
QoS	quality of service
RAN	radio access network
RH	red hat
RMM	resource management module
SaaS	software as a service
SC	smart city
SDNs	software defined networks
TMM	task management module
URM	user registration module

VM	virtual machine
VMDK	virtual machine disk
WAN	wide area network

1. Introduction

Due to the tremendous growth of mobile devices and their usage, the computational demand to execute increased workloads has magnanimously increased. These mobile devices are unable to cope with this increased demand of resources thus requiring Mobile Cloud Computing (MCC). The demand for increased computational resources is addressed by offloading the compute intensive tasks to the resource rich environment of the cloud. However, mobile cloud computing solutions face challenges of latency and low throughput due to increased distance and limited Internet bandwidth between cloud and mobile devices. Edge Computing (EC) solutions such as cloudlet computing, Fog Computing (FC) and Mobile Edge Computing (MEC) are considered more viable as these solutions overcome the challenges of latency and limited Internet bandwidth by placing the computational facility closer to the user, preferably in the same Local Area Network (LAN). The edge devices are backed up with more computational power and stable Internet bandwidth to further offload the task to the cloud if the desired resources are not available at edge nodes. In the current scenario of growth in both Internet of Things (IoT) and Smart City (SC) sectors, edge computing solutions are faced with resource scarcity challenges. Since edge computing solutions do not offer resource sharing and load balancing with neighbor edge devices, an edge node is forced to forward the request to a remote cloud to overcome the limitation of resources. Hence, it can be concluded that edge solutions are facing the same challenges of latency and limited bandwidth as MCC solutions.

These limitations are addressed by a federated cloudlet solution. The federated solution as shown in Fig. 1 comprises different cloudlets and a central broker. The cloudlets share resources with their neighbor cloudlets and hence in case of resource shortage at a certain cloudlet, the request is referred to a neighbor cloudlet instead of a remote cloud. Since the neighbor cloudlet is closer to the source cloudlet as a comrade to the remote cloud, it improves the offloading, request processing, execution, and migration time.

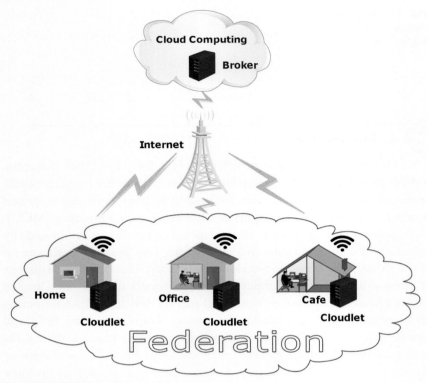

Fig. 1 Cloudlet Federation Model.

In light of the above arguments, we can conclude that the proposed Cloudlet Federation for Resource Optimization (CFRO) model adheres to the qualities of closer proximity and scalability from cloudlet and MCC based models respectively and none of their weaknesses, i.e., latency and limited bandwidth [1]. The tasks such as cloudlet membership, resource management, optimal placement and migration decision are performed by the broker. The broker receives the resource information from all the member cloudlets and pushes this cumulative information back to each member cloudlet from time to time for decision making in the absence of a broker to avoid the central point of failure. A user gets connected with a cloudlet and initiates a request. This request is sent to the broker and the decision about optimal cloudlet for VM placement is received back. An optimal cloudlet is one having adequate resources and offering minimum total delivery time. Total delivery time refers to the offloading, migration and execution time. In case the broker is not available, a cloudlet may take a placement decision using the information matrix pushed by the broker that contains the resource information of all member cloudlets. The optimality of the decision and solution space may vary due to the age of information.

For the integrity of any solution, these solutions must be rigorously tested. Various simulation and emulation platforms are used for the testing of cloud computing solutions among which DevStack, oVirt, Jellyfish, Scalr, and VMWrae Vcenter are famous ones. However, these platforms need to be evaluated for resource sharing and load balancing features for a federated cloudlet environment. The underlying requirements of a cloudlet federation consist of multiple clouds and cloudlets management, resource monitoring and VM migration between cloudlets that may or may not belong to the same cloud. This evaluation produces a need for the survey of existing virtualization platforms highlighting their suitability for a federated cloudlet environment. In case of the unsuitability of these virtualization platforms for a federated cloudlet environment, a novel testbed is required.

The rest of the chapter is organized as follows: Section 2 presents the background and Section 3 details the related work. Section 4 provides the detailed design and architecture of the proposed testbed for federated cloudlet environment and Section 5 concludes the chapter with possible conclusions and future directions.

2. Background

Cloud computing has gained interest recently due to the overwhelming computational requirements of diverse applications. Managing such applications locally requires specialized hardware, software, and network resources. The level of expertise required to manage such complex applications is also high. These all factors increase the cost of the solution thus making it unfeasible for the organizations. On the other hand, cloud computing uses shared resources and provides the same expertise at a much cheaper cost. Multiple Virtual Machines are deployed on a single physical server to maximize resource utilization and reduce cost [2]. The seamless services and ease of use are some other attractive factors of cloud computing. Cloud computing offers different service-oriented models such as IaaS (Infrastructure as a Service), PaaS (Platform as a Service), DaaS (Database as a Service), and SaaS (Software as a Service) to fulfill diverse needs of infrastructure, platform, database and software respectively [3]. The other similar paradigms to cloud computing are Mobile Cloud Computing (MCC) [4] and Edge Computing (EC) [5]. MCC provides resource augmentation for resource limited mobile devices by offloading the compute intensive tasks to the cloud. Edge computing solutions include Mobile Edge Computing (MEC) [6], cloudlet computing [7], and Fog Computing (FC) [8]. Both MCC and edge computing solutions are used to augment the limited resources of different devices [9–11]. A summary of the advantages and disadvantages for both MCC and Edge Computing is presented in Table 1.

In MCC, the compute intensive tasks are offloaded to the remote cloud through the Internet. These tasks are executed in the resource rich environment offered by cloud computing and results are sent back to the user [12,13]. A variety of gaming, social networks and image processing applications are getting advantages from MCC [14–18]. One perspective of mobile cloud computing is to address the challenges of limited power, computational and network resources while the other perspective is to provide business opportunities to the cloud and Internet service providers [19]. This concept is also known as Cloud-based Mobile Augmentation (CMA) [20] or Cyber Foraging [21]. Despite these advantages, users still face

Table 1 Advantages and disadvantages.

Technique	Advantages	Disadvantages
Mobile Cloud Computing	− Better accessibility: Resources are accessible from anywhere through the Internet − Disaster recovery: Less prone to disasters due to redundancy of resources − Diversified features: Support for more features than any other model	− Internet dependency: Cannot work without Internet access − Increased latency: Due to longer distance from source to destination
Edge Computing	− No dependency over the Internet: As most of the communication is on the Local Area Network (LAN) − Closer proximity: One hop distance away from mobile device makes it faster and easier to access − Stable connectivity: Wi-Fi and LAN are used that provide stable connectivity	− Less features: As compared to the remote cloud, cloudlets support a smaller number of features − Number of users limitation: Cloudlets also suffer from number of users limitation as compared to a public cloud

challenges of seamless connectivity between mobile devices and cloud, limited bandwidth and increased latency due to distance between a mobile device and remote cloud. These challenges are addressed by edge based solutions by bringing the computational facility closer to the user.

In MEC the computation facility is placed at the edge of a Radio Access Network (RAN) whereas, Fog computing M2M (Machine-2-Machine) gateways are placed near to the source device. Cloudlets on the other hand are dedicated stations like mini clouds in closer proximity of the user. The term "cloudlet" was first coined by M. Satyanarayanan [22]. These solutions bring the computational facility in closer proximity of the users to avoid latency and limited bandwidth challenges. The concept of fog computing is often used in the Internet of Things (IoT) where either the end device or a central controller device provides the computation facility near to data origin [23]. Delay sensitive applications perform better in an edge computing environment as compared to MCC [24,25]. The application of connected cars can be taken as an example of delay sensitive applications where cars possessing self-driving capabilities have to communicate with other vehicles and road side infrastructure simultaneously to obtain traffic alerts. The existing edge based solution used by this application provides a latency of less than 20 ms. However, this amount of latency still needs improvement and a benchmark of 1 ms has been set for the future generation systems [26]. Real time audio and visual data is another example of delay sensitive application [27–30]. Due to the diversity of applications and their demand for unique features require more flexible solutions such as those provided by cloudlet computing [31].

Cloudlet is a mini cloud in closer proximity having rich computing resources and stable Internet connection [32–36]. Cloudlet is available in the same Local Area Network preferably one hop away from the mobile device. Cloudlets can be fixed or mobile and can also be formed dynamically as per need [37]. Let's take an example of Amazon's EC2 infrastructure to conclude the benefits of cloudlet computing. The instances of Amazon's EC2 are available in only six cities worldwide. A user who wants to offload a compute intensive task has to reach one of these instances through Wide Area Network (WAN). The extended path between a mobile device and Amazon's instance results in increased network latency. The dependency over the Internet and delay due to distance are two important factors that generate the necessity of cloudlet computing [38]. On the other hand, an increased number of devices, IoT sensors, and smart city infrastructure have produced the demand for more computational resources in the vicinity thus making cloudlet computing inadequate in terms of resources. IoT devices [39–42] and smart city Information Communication Technologies (ICTs) [43–45] are two main building blocks of the Information Technology World (ITW).

The concept of federation is used in cloud computing for resource sharing and load balancing between different CSPs for mutual business gain [46]. However, the concept of Cloudlet Federation differs from cloud federation in a way that the federation is formed between different cloudlets that may or may not belong to the same CSP [1]. The ownership of a cloudlet is often private and the coverage is usually limited to a single building, home, office, campus, factory or cafe, etc. The other difference is the closer proximity to a user which is the main objective of edge computing. A unique advantage of the federated cloudlet model includes the availability of extended solution space for the optimal selection of a cloudlet by considering available resources, hop-count, offloading, energy state and migration time, etc. [1,47]. It helps extend the coverage of an individual cloudlet to MAN and WAN environments being a part of the federation.

Various simulators, emulators and virtualization platforms such as DevStack, Scalar, oVirt, Jellyfish and VMWare vCenter can be used to realize the concept of cloud or cloud federation. However, since the requirements of a cloudlet federation differ from a cloud federation, an extensive exploration of the state-of-the-art solutions is therefore required.

3. Related work

This section presents a detailed analysis of the available virtualization platforms. The objective of this analysis is to select and evaluate a suitable environment to execute the federated cloudlet model. The evaluation criterion for the desired testbed is based on the required support for cloudlet federation and includes the core functionalities such as management of multiple clouds, client side resource monitoring and VM migration between cloudlets. Multi-cloud management is the concept of managing multiple cloud service providers together in a collaborative manner. Any service can be provisioned from any CSP. Client side monitoring is the facility to observe the resource status of all member cloudlets. So, optimal selection and load balancing decisions can be taken to optimize resources. VM migration between cloudlets belonging to different CSPs and having different infrastructure is required to form a cloudlet federation.

This analysis includes major virtualization platforms such as VMWare vCenter, DevStack, Scalr, oVirt, and Jellyfish. vCenter is the product of VMWare Inc. offering virtualization platforms to host virtual machines. The salient features of vCenter include ease of deployment and administration, centralized control and visibility, scalability and extensibility. DevStack is the development version of OpenStack produced by the OpenStack community. The features offered by DevStack include orchestration, resource tracking, scheduling, monitoring, graphical user interface, application programming interface, and security. Scalr is an open source enterprise grade cloud management platform developed by Scalr Inc. The functionalities of Scalr includes cost analytics, security, resource tagging, predefined templates, and policy management. oVirt is an open source virtualization platform developed by Red Hat. The features offered by oVirt are easy administration, availability, resource utilization, fast deployment and snapshotting. Jellyfish is another open source platform developed by Jellyfish Inc. which is famous for multi-cloud management. Jellyfish has extensive reporting functionalities and is also mobile ready.

All the aforementioned virtualization platforms are deployed at the Advance Communication lab to assess the availability of required features for the cloudlet federation that includes multi-cloud management, resource monitoring of cloudlets, and VM migration between cloudlets. A detailed discussion and feature wise comparative analysis of these virtualization platforms is presented to conclude the unavailability of any suitable virtualization platform for cloudlet federation.

VMware vCenter Server [48] is a virtualization platform offering a centralized model to manage VMware vSphere environments. VMware vSphere is the product launched by VMWare to manage virtual machines and offer Infrastructure as a Service (IaaS) model. vSphere offers a complete cloud environment offering transparency of physical servers and VMs. This product is enterprise grade and known to offer the best performance for infrastructure and applications. The main features of vCenter Server include ease of deployment and administration, centralized control and visibility, Scalable and Extensible Platform. Ease of deployment and administration means that the Graphical User Interface (GUI) offered by vCenter is user friendly and does not require a very highly trained expert to use it. All the vSphere servers can be easily added at one location to be managed from a central point. However, the current implementation of vCenter is only for the windows server environment. Centralized control and visibility refers to the web interface of vSphere that can be accessed using any supported browser for HTML 5. All kinds of features are available in the web interface as well, for example; create users, assign roles, add inventory, initialize and control VM. Scalability and extensibility refers to the fact that a single vCenter server can be used to manage up to 1000 hosts and 10,000 virtual machines. vCenter server and vSphere are commercial products, and hence, cannot be used for research purposes. Although vSphere is Linux based but the kernel is not open source and hence cannot be modified. Vsphere appliances are also not directly exportable to other virtualization platforms such as offered by Oracle, Ovirt, and OpenStack.

DevStack [49] is a development environment offered by the OpenStack community for researchers and scientists. DevStack has a GUI and Command line Interface (CLI) for configurations. It offers tools for creating and managing a cloud environment. The main features of Devstack include orchestration resource tracking, scheduling, monitoring, GUI, Application Programming Interfaces (APIs), and Access Control Lists (ACLs). Orchestration refers to predefined templates for automated VM initialization and deployments. Resource tracking means that the available resources for the cloud environment are automatically tracked for the provisioning of VMs. Scheduling provides built-in business logic and the process of automatic resource discovery and provisioning yielding better performance. Monitoring is linked to extensive reporting system to determine the resource usage by users. GUI provides easy manageability, deployment and configurations for the end user. The maintenance mode allows users to customize and make changes to their VMs very easily. APIs are available for the integration of any third party solutions. ACLs and role based provisioning is available to protect the identity of the user and to facilitate multi-tenancy. However, DevStack cannot be used in the production environment as it is only for short time testing. It cannot be shut down and restarted like an enterprise system. The installation method is script base and very complex. The installation interface is the command line and does not offer extensive help for troubleshooting. If an installation fails, it has to be started from scratch after explicitly cleaning up the previous incomplete installation. The installation is Internet dependent and takes a lot of time and bandwidth resources. The offline version of DevStack is not available. Only a limited set of built-in VM images are available. The creation of a customized instance and network topology is very complex and requires an expert level of understanding. From version "Pike" VM migration feature has been removed.

Scalr [50] is an open source Cloud Management Platform (CMP) offering multi-cloud infrastructure management. Scalr aims to provide a single

centralized platform to manage multiple cloud environments simultaneously. The objective of Scalr is to enable organizations to manage public,
private, and hybrid cloud environments by using a single graphical user
interface. Administrators can create and remove cloud instances, monitor
cloud resource utilization, automate workflows and assign costs. The main
features of Scalr include cost analytics, security, resource tagging, predefined
templates, and policy management. Cost analytics provides an extensive
reporting to manage the budget by associating the costs with resources
and keeping track of it. Security offers the functionality of a single sign-
on for multiple clouds belonging to a single user. It can also be integrated
with Microsoft Windows active directory or OpenLDAP server for user
authentication and management. Resource tagging lets users add tags to
cloud resources upon launch to help track usage and costs. Predefined
templates gives cloud developers and engineers pre-made templates to
streamline application development. Users can also create self-service cata-
logs. Policy management allows administrators to create and enforce gover-
nance of policies across their cloud infrastructure. Scalr does provide the
facility to manage multiple clouds from a single window but it does not provide client side resource monitoring and VM migration between multi-
cloud environments. These two features are very vital for cloudlet federation
and their absence makes it unsuitable for cloudlet federation.

oVirt [51] is open source platform developed by Red Hat (RH) on
which the Enterprise Virtualization Platform (EVP) is based. It has a web
interface providing ease of use to manage VMs and the underlying infrastructure consisting of storage, compute and network nodes.

The oVirt platform offers various features such as administration,
availability, resource utilization, fast deployment and snapshotting.
Administration refers to the single web interface providing ease of management to handle multiple VMs with complete control of underlying physical
servers. Separate storage and compute nodes can be configured easily working together as a single unit. Existing VMs on other infrastructures can also
be added to the oVirt environment. Availability ensures that the VMs are not
dependent upon a single physical server, these can be migrated to another
available physical server automatically as per load and unavailability of a
particular physical server. Resource utilization provides a customized policy
implementation feature to reduce cloudlet consumption and cost by
automatic load balancing between physical servers. Fast deployment and
snapshotting offers predefined templates for the fast and rapid deployments
of VMs with the automated feature of snapshotting at scheduled times.

Although oVirt is an open source tool but the client Operating System (OS) has a custom kernel that cannot be modified or recompiled limiting its research utilization. oVirt offers live VM migration features between clients but cannot be integrated with the other third party virtualization platforms. It also does not provide the functionality to manage multiple cloud environments.

Project Jellyfish [52] is an open source platform that offers the functionality to manage multiple cloud environments simultaneously. It introduces a centralized brokerage system to mediate between different cloud environments and enables the administrator to manage these clouds with a policy driven approach. The main features of project Jellyfish include multi-cloud management, mobile readiness, and reporting. Multi-cloud management provides allows single user to add all of his/her purchased services from different clouds and associate a specific set of policies to manage resources and costs. Mobile readiness ensures that users can manage their activities and change policies whenever required from a mobile device. Reporting allows users to monitor their resource utilization and wastage to further optimize the rules and policies for cost saving in the future. The focus of jellyfish is to manage multiple clouds from the reporting point of view rather than from the operational point of view which makes it unsuitable for cloudlet federation.

The detailed discussion about aforementioned virtualization platforms leads to the conclusions that there is no open source software that completely fulfills the requirements for resource sharing and load balancing in a way that it can be customized or modified according to the specific needs and used for research purposes. DevStack is a development version of OpenStack and does not offer VM migration. VMWare vCenter does support VM migration but is not an open source software and does not allow any sort of modifications in the code. oVirt supports VM migration and is an open source software but uses a custom kernel and hence cannot be extended or modified. Scalr and Jellyfish do not support VM migration.

This produces a need for a novel testbed to provide the desirable features unavailable in the state-of-the-art for research purposes. The features of customization and VM migration are very vital for a federated environment. The properties of an ideal platform include open source environment, customizable, support and capability to integrate with other virtualization solutions, interdependent of OS, hypervisor and VM disk format.

The proposed ClPyZ is an open source platform developed at Advanced Communication Networks Lab for research purposes. It can manage multiple cloudlets even if they belong to different clouds. A central broker

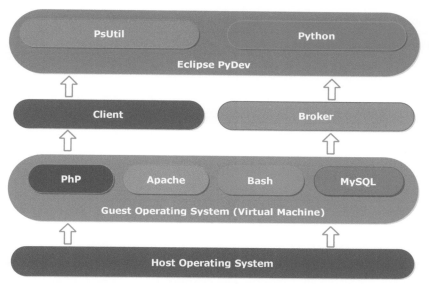

Fig. 2 ClPyZ Design.

manages all the affairs such as cloudlet registration, finding an optimal cloudlet for task offloading and manage all the information of member cloudlets. An architectural layout of ClPyZ has been presented in Fig. 2.

The broker part is built in Python version 2.7 language that uses a combination of PsUtil library with Eclipse PyDev extension to obtain the basic functionality of resource gathering from the operating system. A web interface built in PhP has been provided on client side to take resource request from the users. The request data is sent to broker and a copy is maintained in the local database using MySQL for record keeping. The features offered by ClPyZ are highly customizable as per requirement. Another advantage of ClPyZ is its platform independent nature that makes it equally suitable for heterogeneous cloudlets. It can work with a variety of supported disk formats with the flexibility to add more and can be installed on any UNIX based operating system. The core functionalities and features offered by ClPyZ include GUI for clients, custom orchestration, multi-cloud management, monitoring, resource tracking, VM migration, centralized control and visibility. GUI for clients provides a dashboard for clients to manage and monitor their requests. Custom orchestration refers to a VM in Open Virtualization Format (OVF) being used in any hypervisor at user's end can be easily ported into the system. Multi-cloud management means that cloudlet federation provides the facility to add multiple cloudlets as members that may or may not belong to same clouds. Live monitoring of resources is

Table 2 Comparative analysis of features for resource federation supported tools.

Features	Cloud Registration	–	✓	✓	–	✓	✓
	Resource Management	✓	✓	✓	✓	–	✓
	Information Management	✓	✓	✓	✓	✓	✓
	Migration	–	✓	–	✓	–	✓
	Resource Monitor	✓	✓	–	✓	–	✓
	Manage Multiple Clouds	–	✓	✓	–	✓	✓
Platforms		DevStack	VMWare vCenter	Scalr	oVirt	Jellyfish	ClPyZ

available to prevent any under provision or overprovisioning of resources with an automated alert system. The resource tracker keeps an eye on the new and lost resources to maintain an updated picture of the resource pool available to the federation. Centralized control and visibility provides server side console to facilitate user by bringing the control and management of all member cloudlets at one place. VM Migration provides the functionality to migrate a VM from one cloudlet to another due to resource constraint or to bring it in the closer proximity of the user to improve the performance.

A comparative analysis of the features offered by the state-of-the-art virtualization platforms mentioned and the proposed has been presented in Table 2.

4. Testbed overview

The proposed testbed as shown in Fig. 3 has two parts, i.e., Broker and Cloudlet. The broker has three modules, i.e., (i) Cloudlet Registration Module (CRM) (ii) Decision Support System (DSS) and (iii) Information Management Module (IMM). Cloudlet has three main modules, i.e., (i) Resource Management Module (RMM) (ii) Task Management Module (TMM) (iii) User Registration Module (URM) and an auxiliary module, i.e., broker-agent acting on behalf of the broker.

Each module consists of various components and services that are responsible for intra-module communication. A module manager is available to manage inter-module operations. The detail of all the modules is given in the subsequent sections.

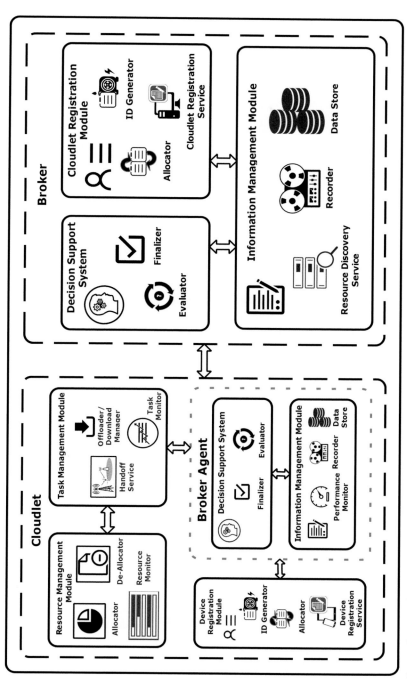

Fig. 3 CIPyZ Architecture.

4.1 Cloudlet module

Cloudlet is a mini cloud in the vicinity of the user providing services to the mobile devices and comprises of the following three modules with an auxiliary broker–agent.

4.1.1 User registration module

User Registration Module is responsible for the registration of a mobile user as shown in Fig. 4.

After registration, a mobile user may send a task offloading request. A task offloading request contains request information (user id, required resources) and data (code, application or VM). A request is considered complete, if all the phases of offloading, execution, and result returning are successfully executed. An incomplete offloading request means, that at least one of these phases was not successful.

User Registration Module has the following four components.

(a) **User Registration Service:** This service is responsible for registering a mobile user and sending the same information to IMM.

(b) **Request Processor:** The request Processor receives the request for registration containing information such as user id and passwords.

(c) **ID Generator:** ID Generator is responsible for generating unique IDs for mobile users.

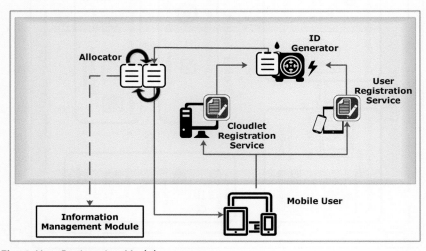

Fig. 4 User Registration Module.

(d) **Allocator:** Allocator is responsible for assigning these IDs to mobile users and sending the same information to IMM.

4.1.2 Task management module

This module in Fig. 5 is responsible for offloading a task to the cloudlet. Offloading a task means transferring a task and relevant data required to execute the task. It can be transferred in the form of code, application or VM. Task Management Module has the following three components.

(a) **Offloader/Download Manager (DM):** This component receives the task and relevant data from a mobile device and offloads it to the cloudlet.

(b) **Handoff Service:** This component is responsible for saving VM state and relevant data to migrate/offload to the optimal cloudlet.

(c) **Task Monitor:** This component continuously monitors different phases of a task such as initiation, request receiving, task offloading, task execution and result returning. It also communicates the status of each phase as successful/unsuccessful to the broker with cloudlet ID and user ID to keep track of request status.

4.1.3 Resource management module

This module in Fig. 6 is responsible for managing resources. When optimal cloudlet is finalized, the resource manager requests for the reservation of resources to offload and execute a task. After the completion of the task, these allocated resources are released and added to the pool of available resources.

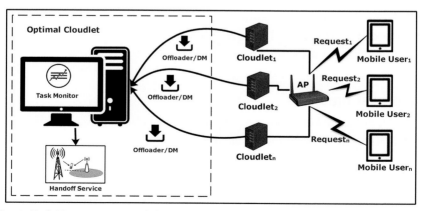

Fig. 5 Task Management Module.

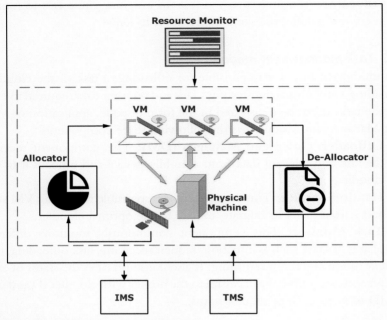

Fig. 6 Resource Management Module.

Resource Management Module has the following three components

(a) **Allocator:** This component receives the required resource information from the information management module and requests the destination cloudlet to reserve the required resources.

(b) **De-Allocator:** It receives the completion status from the task management module and de-allocates the occupied resources and adds them to
the pool of available resources.

(c) **Resource Monitor:** This is responsible for tracking the resources and
providing resource information to IMM.

4.2 Broker module

Broker is the central entity on a third party remote cloud managing all cloudlet federation operations and comprises of the following three modules.

4.2.1 Cloudlet registration module

The Cloudlet Registration Module is responsible for the registration of
cloudlets as shown in Fig. 7.

Cloudlet Registration Module has the following four components

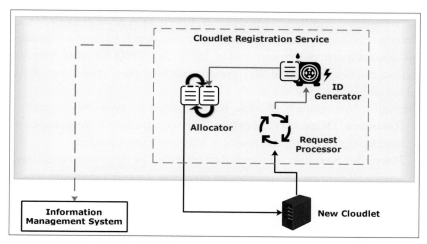

Fig. 7 Cloudlet Registration Module.

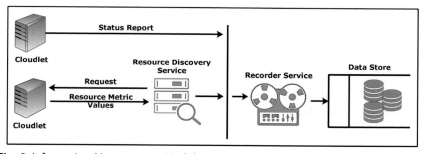

Fig. 8 Information Management Module.

(a) **Cloudlet Registration Service:** This service is responsible for cloudlet registration and sending the same information to Information Management Module.

(b) **Request Processor:** Request Processor receives the request for registration containing hostname, IP address, and owner information.

(c) **ID Generator:** ID Generator is responsible for generating unique IDs for cloudlets.

(d) **Allocator:** Allocator is responsible for assigning IDs to cloudlets.

4.2.2 Information management module

The Information Management Module (IMM) in Fig. 8 is responsible for maintaining information such as resource matrices, optimal selection matrices, status reports and registration information. Information Management Module has two service components and a data store for storing all kinds of information.

Resource matrices are presented in Table 2(a). Table 2(b) shows optimal selection metrics and Table 2(c) presents status reports consisting of flags used for displaying different states.

Information Management Module has the following two components

(a) **Resource Discovery Service:** This component is responsible for requesting/receiving the resource matrices from cloudlets.

(b) **Recorder Service:** It records data provided by the resource discovery service. Recorder service also receives status reports containing status flags indicating successful/unsuccessful operations from cloudlets.

4.2.3 Broker decision support system

This module, as shown in Fig. 9 is responsible for the evaluation and finalization of the optimal cloudlet.

Decision Support System has the following two components.

(a) **Evaluation Service:** This component receives the resource related information from IMM regarding resources at cloudlets and required resources to offload and execute the task. It filters the eligible cloudlets having adequate resources for the requested job and sends this information to the finalizer service.

(b) **Finalizer Service:** This component receives eligible cloudlets' list to finalize the selection of optimal cloudlet. An optimal cloudlet is one having available resources with minimum total delivery time.

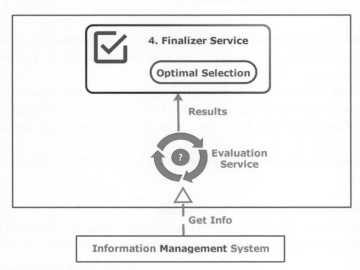

Fig. 9 Decision Support System.

4.3 Broker-agent

Broker-agent works on behalf of the broker on a cloudlet and comprises the Information Management Module and Decision Support System. In case the broker is not available, the decision support system of broker-agent takes the optimal decision based on the information stored in its information management module.

5. Implementation

5.1 Sample scenario

A sample scenario is created consisting of one broker and six cloudlets. Three different locations have been selected in the same city to host the cloudlets. Two cloudlets at each location are hosted using two different Internet Service Providers (ISP) to mimic both Metropolitan Area Network (MAN) and Wide Area Network (WAN) environments (Fig. 10).

The broker is hosted on Amazon's EC2 cluster to mimic a remote cloud while the cloudlet nodes are hosted in a local datacenter. The details of the testbed are presented in Table 3.

A physical server using the following specifications as mentioned in Table 4 has been used to host the virtual cloudlet nodes each configured with

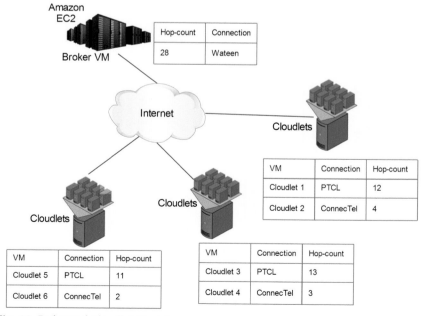

Fig. 10 Federated Cloudlet Model.

Table 3 Information description of IMM (broker).

(a) Resource matrices

Parameters				Matrix	Description
CPU	M	B	DS	ARM	Available resources on the cloudlet
				RRM	Required resource to offload and execute the request
Data Size		B		ORM	Offloaded data size on cloudlets and available bandwidth from source cloudlet to optimal cloudlet
M = Memory					RRM = Required Resource Matrix
B = Bandwidth					ORM = Offloading Resource Matrix
				DS = Disk Space	
				ARM = Available Resource Matrix	

(b) Optimal selection matrices

Parameters				Matrix		Description
JQT	ET		OT	HC	OSMC	Contains parameter values to select optimal cloudlet.
JQT = Job Queue Time						ET = Execution Time
HC = Hop Count						OT = Offloading Time
						OSMC = Optimal Selection Matrix for Cloudlet

(c) Description of flags used for status reporting

Flags	Description
TT	Successful/unsuccessful transfer of task from source cloudlet to optimal cloudlet
OT	Successful/unsuccessful offloading of a task from mobile device to immediate cloudlet
TE	Successful/unsuccessful completion of task execution phase
TRIC	Successful/unsuccessful transfer of result to immediate cloudlet
TRMD	Successful/unsuccessful transfer of result to mobile device
TT = Transfer Task	OT = Offloading Task	TRMD = Transfer Result to Mobile Device
TE = Task Execution	TRIC = Transfer Result to Immediate Cloudlet	

Table 4 Testbed details.

	Nodes	Locations	ISPs
Broker	1	1	Wateen Pvt. Ltd
Cloudlet	2	2	PTCL/ConnecTel
Cloudlet	2	3	PTCL/ConnecTel
Cloudlet	2	4	PTCL/ConnecTel

Table 5 Physical and virtual specifications.

	Model	CPU Cores	Frequency	Type	RAM	HDD	NIC	BW
Server	HP DL360	4	2.8GHZ	Xeon	64 GB	250 GB	1 Gbps	32 Mbps
Guest VMs	ESX 6.0	1	–	–	8 GB	30 GB	–	–
User VM	Tiny Core Linux	1	–	–	48 MB	45 MB	–	–

the following specifications as mentioned in Table 4. The operating system used in each VM is Ubuntu 14.04 Long Term Support (LTS).

The request flow is initiated by a registered user from a cloudlet. The user provides resource requirements for the VM to be offloaded to the cloudlet and submits a request. A very small VM having a size of 14.7 MB on disk in Open Virtualization Appliance (OVA) file format is used in the experimentations. Further specifications of the user VM are provided in Table 4. The immediate cloudlet with which the user is currently connected forwards this request to the broker. The broker then finalizes the placement decision by evaluating an optimal cloudlet for the given request and these results are pushed back to the respective cloudlet. In our model, since the request and decision both consist of only the information matrix and not the actual VM, hence their latency is negligible. Sample metric values for experimental setups are shown in Table 5.

The decision contains the target cloudlet's address where the request should be executed. It can be the same cloudlet with which the user is currently connected and which forwarded the request to broker or any other in the federation. The execution place of the request has more impact on the latency as the large VM file has to be transferred at the target cloudlet.

The sample scenario contains three different experimental setups, one with a distant remote server using Amazon EC2 instances and is labeled as Amazon, the second using the same ISP mimicking a MAN environment and is labeled as SISP, and the third using different ISPs mimicking a WAN environment and is labeled as DISP. The detailed configuration for each experimental setup have already been provided in Section 4.3.

5.2 Performance metrics

All the following metrics have been measured between source to destination cloudlets where source refers to the cloudlet forwarding the request to broker and destination refers to the target cloudlet where the request is to be executed. The first metric is latency l, second is throughput τ, the third one is hop-count h, and the fourth is migration time t_m.

5.3 Results

The results of each metric have been evaluated and presented in the form of histograms. All the experiments are conducted in an isolated production environment. Each experiment has been repeated several times to obtain values of various performance and resource parameters. The trials are selected from four different experiments at random times to observe the effects of changes in the network topology (Table 6).

5.3.1 Latency and hop-count

Distance refers to the length of the communication path between source and destination and hop-count refers to the number of intermediate devices. A hop-count is often referred to as the number of routers on the path from the source to destination. Latency l has a direct relationship with distance d and hop-count h. Each hop-count incurs delay due to the processing of packets. The results for latency have been presented in Fig. 11 and the results for hop-count are presented in Fig. 12. The x-axis of the latency and hop-count graphs represents randomly selected trials and the y-axis shows latency in milliseconds and hop-count respectively. The latency results

Table 6 Sample metric values for experimentation.

Study Case	1	2	3	4	5
File Size	14.7 MB	14.7 MB	14.7 MB	14.7 MB	14.7 MB
Number of Requests	15	30	45	60	75
Users	1	5	10	20	25
Bandwidth	32 Mbps up/down				

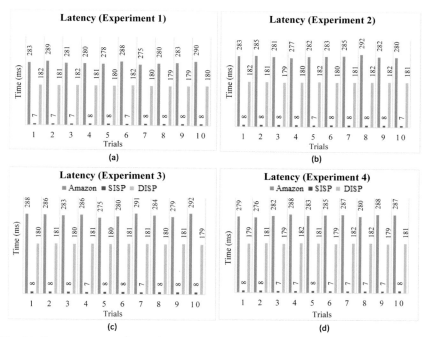

Fig. 11 Comparative Analysis of Latency.

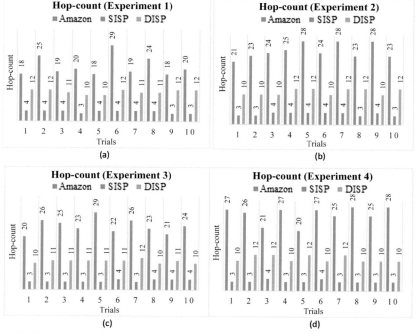

Fig. 12 Comparative Analysis of Hop-count.

presented in Fig. 11 clearly show that the lower latency values can be achieved through the same ISP setup, moderate latency values can be obtained using different ISP setups, while higher latency values are seen with distant remote server setup using Amazon EC2 instances. The hop-count results presented in Fig. 12 clearly show that the lower hop-count values can be achieved through the same ISP setup, moderate hop-count values can be obtained using different ISP setup, while higher hop-count values are seen with distant remote server setup using Amazon EC2 instances.

5.3.2 *Migration time and throughput*

Migration time refers to the time required for a VM to migrate from one server to another. Migration time is highly dependent upon VM file size and throughput. The migration time t_m has a direct relationship with VM file size S_f and an inverse relationship with throughout τ. The results for migration time has been presented in Fig. 13 and results for throughput are presented in Fig. 14. The x-axis of the migration time and throughput

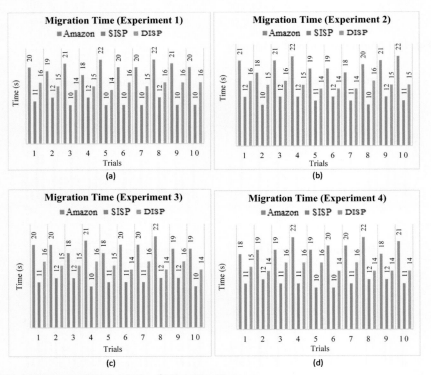

Fig. 13 Comparative Analysis of Migration Time.

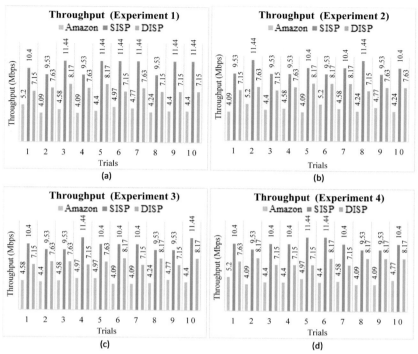

Fig. 14 Comparative Analysis of Throughput.

graphs represents random selected trials and the y-axis shows time in milli-seconds and throughput in Megabits per seconds (Mbps) respectively. The migration time results presented in Fig. 13 clearly show that the lower migration time can be achieved through the same ISP setup, moderate migration time can be obtained using different ISP setup, while higher migration time is seen with distant remote server setup using Amazon EC2 instances. The throughput results presented in Fig. 14 clearly show that the higher throughput can be achieved through the same ISP setup, moderate throughput can be obtained using different ISP setup, while lower throughput values are seen with distant remote server setup using Amazon EC2 instances.

6. Conclusions and future directions

It can be concluded from the above findings that a new testbed is needed to mimic the federated cloudlet environment offering features of VM migration, resource collaboration, and optimal cloudlet selection.

These features are available in the proposed testbed ClPyZ. The major design considerations of the proposed testbed include the management of multiple cloudlets even if they belong to different clouds. A central broker manages all the affairs such as cloudlet registration, finding an optimal cloudlet for task offloading and managing all the information of member cloudlets. A request is referred to the broker and the final decision is pushed back to the source cloudlet. Similarly, the broker keeps track of the running tasks and resources occupied by them. So, in case a cloudlet is short of resources, the load balancing mechanism can be initiated by migrating a VM from a resource constrained cloudlet to a resource available cloudlet. ClPyZ is ready for any production environment with the features of scalability, improved Quality of Service (QoS), and Quality of Experience (QoE). The generic features and independent design enable ClPyZ to be integrated with other proprietary solutions such as DevStack, VMWare ESX, Amazon EC2 and Google compute engine.

The future research work will be shaped by keeping the possible enhancements and practical implications in perspective. Following are some possible enhancements and practical implications that can be added to the proposed testbed. ClPyZ currently supports OVA files compatible with Oracle Virtual Box (OVB). Enhancement for other formats such as Open Virtualization Appliance (OVA) and Virtual Machine Disk (VMDK) can be added for more flexibility. Extensive testing for the production environment with full sized VM having standard OS can be performed to better analyze the efficacy of the solution. Simultaneous migrations generate a lot of network traffic between source and destination cloudlets that may affect other traffic. Using Software Defined Networks (SDNs) and multipath circuits, the traffic can be rerouted dynamically avoiding network congestion. The system-level commands have been used to fetch the hardware information that varies for different operating systems and hence are required to be changed according to the OS acceptable commands. A pre-configuration step can be added to change the commands according to the specific OS to address this limitation.

References

[1] M.Z. Nayyer, I. Raza, S.A. Hussain, CFRO: cloudlet federation for resource optimization, IEEE Access 8 (2020) 106234–106246.
[2] M.Z. Nayyera, I. Razab, S.A. Hussainb, Revisiting VM performance and optimization challenges for big data, Adv. Comput. 114 (2019) 71.
[3] M.R. Rahimi, et al., Mobile cloud computing: a survey, state of art and future directions, Mob. Netw. Appl. 19 (2) (2014) 133–143.

[4] H. Allam, et al., A critical overview of latest challenges and solutions of mobile cloud computing, in: Fog and Mobile Edge Computing (FMEC), 2017 Second International Conference on, IEEE, 2017.

[5] M. Satyanarayanan, The emergence of edge computing, Computer 50 (1) (2017) 30–39.

[6] N. Abbas, et al., Mobile edge computing: a survey, IEEE Internet Things J. 5 (1) (2018) 450–465.

[7] Z. Pang, et al., A survey of cloudlet based mobile computing, in: Cloud Computing and Big Data (CCBD), 2015 International Conference on, IEEE, 2015.

[8] R. Mahmud, R. Kotagiri, R. Buyya, Fog computing: a taxonomy, survey and future directions, in: Internet of Everything, Springer, 2018, pp. 103–130.

[9] K. Dolui, S.K. Datta, Comparison of edge computing implementations: fog computing, cloudlet and mobile edge computing, in: Global Internet of Things Summit (GIoTS), 2017, pp. 1–6.

[10] P. Mach, Z. Becvar, Mobile edge computing: a survey on architecture and computation offloading, IEEE Commun. Surv. Tutor. 19 (3) (2017) 1628–1656.

[11] K. Bilal, et al., Potentials, trends, and prospects in edge technologies: fog, cloudlet, mobile edge, and micro data centers, Comput. Netw. 130 (2018) 94–120.

[12] S. Abolfazli, et al., Cloud-based augmentation for mobile devices: motivation, taxonomies, and open challenges, IEEE Commun. Surv. Tutor. 16 (1) (2014) 337–368.

[13] M. Satyanarayanan, A brief history of cloud offload: a personal journey from odyssey through cyber foraging to cloudlets, GetMobile: Mobile Comput. Commun. 18 (4) (2015) 19–23.

[14] E. Ahmed, et al., Seamless application execution in mobile cloud computing: motivation, taxonomy, and open challenges, J. Netw. Comput. Appl. 52 (2015) 154–172.

[15] N. Aminzadeh, Z. Sanaei, S.H. Ab Hamid, Mobile storage augmentation in mobile cloud computing: taxonomy, approaches, and open issues, Simul. Model. Pract. Theory 50 (2015) 96–108.

[16] S. Distefano, F. Longo, M. Scarpa, QoS assessment of mobile crowdsensing services, J. Grid Comput. 13 (4) (2015) 629–650.

[17] M. Othman, S.A. Madani, S.U. Khan, A survey of mobile cloud computing application models, IEEE Commun. Surv. Tutor. 16 (1) (2014) 393–413.

[18] M. Othman, et al., MobiByte: an application development model for mobile cloud computing, J. Grid Comput. 13 (4) (2015) 605–628.

[19] H. Qi, A. Gani, Research on mobile cloud computing: review, trend and perspectives, in: Digital Information and Communication Technology and it's Applications (DICTAP), 2012 Second International Conference on, IEEE, 2012.

[20] A. Mohammad, L. Chunlin, Cloud-based mobile augmentation in mobile cloud computing, Int. J. Future Gener. Commun. Netw. 9 (8) (2016) 65–76.

[21] P. Patil, A. Hakiri, A. Gokhale, Cyber foraging and offloading framework for internet of things, in: Computer Software and Applications Conference (COMPSAC), 2016 IEEE 40th Annual, IEEE, 2016.

[22] M. Satyanarayanan, et al., The case for vm-based cloudlets in mobile computing, IEEE pervasive Computing 8 (4) (2009) 14–23.

[23] S. Yi, C. Li, Q. Li, A survey of fog computing: concepts, applications and issues, in: Proceedings of the 2015 Workshop on Mobile Big Data, 2015, pp. 37–42.

[24] Y.N. Krishnan, C.N. Bhagwat, A.P. Utpat, Fog computing—network based cloud computing, in: Electronics and Communication Systems (ICECS), 2015 2nd International Conference on, IEEE, 2015.

[25] E. Ahmed, M.H. Rehmani, Mobile Edge Computing: Opportunities, Solutions, and Challenges, Elsevier, 2017, pp. 59–63.

[26] D. Sabella, et al., Mobile-edge computing architecture: the role of MEC in the internet of things, IEEE Consum. Electron. Mag. 5 (4) (2016) 84–91.

[27] I. Stojmenovic, S. Wen, The fog computing paradigm: scenarios and security issues, in: Computer Science and Information Systems (FedCSIS), 2014 Federated Conference on, IEEE, 2014.

[28] I. Stojmenovic, Fog computing: a cloud to the ground support for smart things and machine-to-machine networks, in: Telecommunication Networks and Applications Conference (ATNAC), 2014 Australasian, IEEE, 2014.

[29] F. Bonomi, et al., Fog computing and its role in the internet of things, in: Proceedings of the first edition of the MCC workshop on Mobile cloud computing, ACM, 2012.

[30] T. Taleb, et al., Mobile edge computing potential in making cities smarter, IEEE Commun. Mag. 55 (3) (2017) 38–43.

[31] P. Mell, T. Grance, The NIST Definition of Cloud Computing, 800, NIST Special Publication, 2011, p. 145.

[32] K. Gai, et al., Dynamic energy-aware cloudlet-based mobile cloud computing model for green computing, J. Netw. Comput. Appl. 59 (2016) 46–54.

[33] K. Ha, et al., Just-in-time provisioning for cyber foraging, in: Proceeding of the 11th Annual International Conference on Mobile Systems, Applications, and Services, ACM, 2013.

[34] S. Simanta, et al., A reference architecture for mobile code offload in hostile environments, in: International Conference on Mobile Computing, Applications, and Services, Springer, 2012.

[35] Y. Wu, L. Ying, A cloudlet-based multi-lateral resource exchange framework for mobile users, in: Computer Communications (INFOCOM), 2015 IEEE Conference on, IEEE, 2015.

[36] Y. Zhang, D. Niyato, P. Wang, Offloading in mobile cloudlet systems with intermittent connectivity, IEEE Trans Mob. Comput. 14 (12) (2015) 2516–2529.

[37] A. Bahtovski, M. Gusev, Cloudlet challenges, Proc. Eng. 69 (2014) 704–711.

[38] Y. Gao, et al., Are Cloudlets Necessary?, School of Computer Science, Carnegie Mellon University, Pittsburgh, PA, 2015.

[39] F. Xia, et al., Internet of things, Int. J. Commun. Syst. 25 (9) (2012) 1101.

[40] H. Kopetz, Real-Time Systems: Design Principles for Distributed Embedded Applications, Springer Science & Business Media, 2011.

[41] L. Atzori, A. Iera, G. Morabito, The internet of things: a survey, Comput. Netw 54 (15) (2010) 2787–2805.

[42] J. Gubbi, et al., Internet of things (IoT): a vision, architectural elements, and future directions, Future Gener. Comput. Syst. 29 (7) (2013) 1645–1660.

[43] T. Nam, T.A. Pardo, Conceptualizing smart city with dimensions of technology, people, and institutions, in: Proceedings of the 12th Annual International Digital Government Research Conference: Digital Government Innovation in Challenging Times, ACM, 2011.

[44] A. Cocchia, Smart and digital city: a systematic literature review, in: Smart City, Springer, 2014, pp. 13–43.

[45] K. Su, J. Li, H. Fu, Smart city and the applications, in: Electronics, Communications and Control (ICECC), 2011 International Conference on, IEEE, 2011.

[46] T. Kurze, et al., Cloud federation. Cloud, Comput. Secur. 2011 (2011) 32–38.

[47] M. Inam, M.Z. Nayyer, Energy-aware load balancing in a cloudlet federation, Eng. Proc. 12 (1) (2021) 27.

[48] C. Bunch, Automating VSphere: With VMware VCenter Orchestrator, Prentice Hall Press, 2012.

[49] A. Awasthi, P.R. Gupta, Comparison of openstack installers, Int. J. Innov. Sci. Eng. Technol. 2 (9) (2015).

[50] H. Work, Scalr: the auto-scaling open-source amazon EC2 effort, TechCrunch 3 (2008) 1–6. posted Apr.

[51] R.J. Moorhead, et al., Oceanographic visualization interactive research tool (OVIRT), in: Visual Data Exploration and Analysis, International Society for Optics and Photonics, 1994.
[52] B.A. Hamilton, 2014. https://projectjellyfish.org/.

About the authors

Muhammad Ziad Nayyer is serving as Assistant Professor in Department of Computer Science GIFT University, Gujranwala, Pakistan. He holds a PhD degree in Computer Science from COMSATS University Islamabad, Lahore Campus. He is an active member of Advanced Communication Networks Lab. He has numerous publications on his account including impact factor journal publications and book chapters. He received his MS degree in Computer Science from Government College University (GCU), Lahore, Pakistan in 2011. His area of interests includes Cloud Computing, VM Migration, Mobile Cloud Computing, Cloud Federation, Mobile Edge Computing, Fog Computing, and Cloudlet Computing.

Imran Raza is working as an Assistant Professor in the Department of Computer Science, COMSATS University Islamabad, Lahore Campus, since 2003. He holds BS (CS) and MPhil Computer Science degrees from Pakistan. His areas of interests include Cloud Computing, Mobile Edge Computing, SDN, NFV, Wireless Sensor Networks, MANETS, QoS issue in Networks, and Routing protocols. He has authored and co-authored more than 40 Journal and conference papers. He has been actively involved in simulating CERN O2/FLP upgrades. He has supervised and co-supervised many funded projects related to ICT in Healthcare. He has been member of IEEE and ACM.

Syed Asad Hussain is working as a professor since 2010 and dean faculty of information sciences and technology at COMSATS University Islamabad Pakistan since 2015. He has served as head of computer science department COMSATS University Islamabad Lahore campus from 2008 to 2010 and from 2011 to 2017. He is currently leading communications networks research group at the department of computer science COMSATS University Islamabad Lahore campus since 2005. He obtained his master's degree from Punjab University Lahore, Pakistan, then he proceeded to the UK for his master's from University of Wales, Cardiff, UK. He obtained his PhD from the Queen's University Belfast, UK. He was funded for his PhD by Nortel Networks UK. He was awarded prestigious Endeavour research fellowship by Australian Government for his post doctorate at the University of Sydney, Australia in 2010. He has taught at Queen's University Belfast UK, Lahore University of Management Sciences (LUMS) and University of the Punjab, Pakistan. Currently he is supervising PhD students at COMSATS University and split-site PhD students at Lancaster University, UK in the areas of cloud computing and cybersecurity. Professor Hussain is actively involved in collaborative research with CERN in Switzerland and different universities of the world such as Cardiff University UK, Lancaster University UK, Dalhousie University, Canada and Charles Sturt University, Australia. He is supervising funded projects as principal investigator in the field of healthcare systems. He has authored and co-authored more than 85 journal and conference papers and has written a book titled Active and Programmable Networks for Adaptive Architectures and Services published by Taylor and Francis USA. He has been a member of IEEE. He regularly reviews international journals and conferences papers.

CHAPTER FOUR

The multicore architecture

Tim Guertin[a] and Ali Hurson[b]

[a]Computer Engineering, Missouri University of Science and Technology, Rolla, MO, United States
[b]Department of Electrical and Computer Engineering, Missouri University of Science and Technology, Rolla, MO, United States

Contents

Abstract

The multicore architecture has played a significant role in computer performance improvements since it was first introduced in the early 2000s [2]. It provides performance improvements due to multiple cores executing instructions concurrently supporting both instruction level parallelism and thread level parallelism. The performance improvement can be achieved at lower clock frequencies as compared to superscalar architectures, resulting in higher performance per Watt. Finally, multicore processors provide an advantage over multiprocessor systems as resources can be shared among the cores that would be duplicated on a multiprocessor system.

The advantages of multicore architectures come at the expense of several challenges such as cache coherency and communication among the cores. This chapter is intended to address these architectural challenges and their potential solutions within the scope of the multicore architecture.

Multicore architectures can be heterogeneous or homogeneous. In homogeneous architectures, as the name suggests, all the cores on the device are the same. In heterogeneous architectures, two or more cores on the device are different. Applications may benefit from different combinations of complex and simple cores. Different cores being used for different purposes in the design may provide a more efficient design. The benefits of each type of architecture will be articulated in this article.

Cache coherency is an important challenge in the multicore architecture. One or more levels of cache are private to each core and one or more levels of cache are shared among the cores. This presents a cache coherency challenge that must be managed. Data that may be used by multiple cores may be stored or modified in one core's private cache. Multiple protocols exist to resolve this issue. Different methods will be presented and compared.

As the number of cores has increased, the interconnection framework has become a bottleneck in the system. Communication is necessary between the cores cache memory to ensure cache coherence. This interconnection architecture was traditionally done using a bus. As the number of cores has increased, a bus is not able to efficiently support the traffic and is limiting the performance improvement obtained by adding cores. The Network on a Chip (NoC) architecture has shown to be a more efficient interconnection mechanism that is able to provide performance improvements as the number of cores is increased. The interconnection mechanism among the cores will be discussed.

Finally, to take advantage of the concurrency available from the multicore architecture, software must be developed for parallel execution. Developing and implementing software for parallel execution is much different than developing software for sequential execution. The tools and techniques used to write parallel software will also be discussed.

An overview of the multicore architecture is first discussed. Design issues encountered with the multicore architecture such as cache coherency, interconnection frameworks, and designing software for parallel execution are then examined.

1. Introduction

By the 1990s, computer architects converged to the decision that increasing operating frequency, deep pipelining, and out of order execution are viable approaches in the design and implementation of high-speed computers. This conclusion was manifested in the design and implementation of the superscalar architecture as the architectural backbone of the so-called supercomputer [1]. Though a significant breakthrough, however, power consumption is directly related to the operational frequency and soon it became evident that power consumption and thermal management issues would be major hurdles in the design and implementation of high-speed computers. Hence, the motivation for the multicore architecture.

The multicore architecture provides a method of increasing performance while maintaining or reducing power consumption. Multiple cores can operate at slower clock frequencies, offering improved overall performance due to the parallel execution of application programs while offering better performance per Watt. Consequently, the multicore platform became the viable alternative for the design and implementation of high-speed computers [2]. This chapter is intended to discuss the advantages and challenges within the framework of the multicore configuration.

With a multicore architecture, multiple processing cores (either heterogeneous or homogeneous) are located on a single chip. Each core contains private cache memory, as well as access to a shared cache. Architectural features that must be determined for a multicore design include the number of cores, heterogeneity/homogeneity of the cores, cache memory sharing and coherency, and the interactions among the cores. These design decisions present tradeoffs in performance, cost, and complexity of the design. As the number of cores increases, the opportunity for parallel execution of application programs also increases. At the same time, the complexity of the interconnection architecture and cache coherency management increases. Shared cache memory can provide better cache utilization but increase the effort to maintain coherency. The tradeoffs in these design decisions will be discussed.

The multicore architecture has also raised many new challenges for software designers. Software has been traditionally developed for serial execution. Software developers are traditionally trained to develop sequential software. Software must be developed and written to take advantage of the potential parallelism the multicore architecture provides. Otherwise, full performance improvement capabilities of the hardware cannot be realized. Software parallelism can be created automatically by a compiler which generates machine code from the high-level software language. Parallelism can also be achieved by specifying the parallelism in the high-level language code. These two methods will be addressed.

2. History of the multicore architecture

Single core processors continued to make performance improvements with each generation for many years. The operating frequencies continued to increase to improve performance. Increasing the operating frequency also increases the dynamic power consumption and heat generation [2].

In addition to increasing operating frequencies, the size of transistors were decreasing due to the advances in technology, facilitating more on-chip functionality [2]. This led to leakage current, consuming a significant amount of power. As transistor sizes decrease, the insulation between the source and drain allows leakage current through the transistor when it is in the off state [3]. See Eqs. (1) and (2). The power consumption and thermal management issues have provided significant challenges in improving performance of single core processor [2].

$$P = P_{\text{dynamic}} + P_{\text{static}} \qquad (1)$$

where:

$$P_{\text{static}} = V^2\, I_{\text{leakage}} \text{ and } P_{\text{dynamic}} = A\,C\,V^2 f \qquad (2)$$

C is capacitance, V is the voltage, I_{leakage} is the leakage current, f is the operating frequency, and A is an activity factor related to the amount of switching activity.

The multicore architecture provides a different path to improving performance. With the multicore architecture, performance improvements can be achieved by increasing the number of cores allowing exploitation of parallelism in the application programs without the need to increase operational frequency. As a result, the rate of power consumption growth has decreased as operating frequencies no longer increases at as fast a rate as they were previously with single core processors [4].

The first dual core processor chip was developed by IBM (i.e., IBM Power4) [5] which offered a huge performance increase over its predecessor, the Power3 [5]. Table 1 provides several examples of multicore processors

Table 1 Some commercial multicore processors.

Processor	Year released	Number of cores	Frequency	Power consumption
AMD Athlon [6]	2005	2	2.8 GHz	89 W
Intel Pentium D 925 [7]	2006	2	3.0 GHz	95 W
Core 2 Quad Q6600 [8]	2007	4	2.4 GHz	105 W
AMD Phenom 2 X4x [9]	2009	4	3.4 GHz	125 W
Intel i5 [10]	2009	4	2.66 GHz	95 W
AMD Phenom 2 X6 [11]	2010	6	3.2 GHz	125 W
Intel Xeon Processor E5-2640 v3 [12]	2014	8	2.6 GHz	90 W
AMD Ryzen 7 1700x [13]	2017	8	3.4 GHz	95 W
Intel Core i9-7900X [14]	2017	10	3.3 GHz	140 W
AMD Ryzen Threadripper 1950X [15]	2017	16	3.4 GHz	180 W
Intel Core i9-7980K [16]	2017	18	2.6 GHz	165 W

that have been released over the years. As is evident from the table, the number of cores has continued to increase. Please note that the power consumption has also increased, but at a lower rate as compared to the number of cores.

3. Multicore architectures

An architecture is a multicore architecture if it has two or more cores on a single chip. This is different from a multiprocessor architecture which is a collection of two or more processors on separate chips. Putting the cores on the same chip allows shorter traces in the communication framework between the processors and shared memory. Shorter traces allow operating at faster speeds without the degradation that occurs with longer traces operating at high speeds. It also uses less power to communicate while taking up less space. The cores share resources on the chip that would be otherwise duplicated on a multiprocessor system providing greater resource utilization. These are the principal advantages of a multicore platform over a multiprocessor architecture.

A basic multicore functional architecture consisting of two cores each with a local cache is shown in Fig. 1. The cache memory could be a unified/dedicated and/or a single level/multiple level cache. An interconnection network is used to interface each core's local cache memory to the

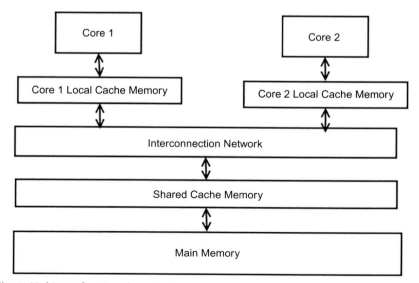

Fig. 1 Multicore functional architecture.

shared cache memory. The shared cache memory may include multiple levels or only one depending on the architecture. The shared cache memory interfaces with the main memory.

Since its inception, as it is witnessed in the literature, many different multicore architectures have been designed and produced (see Table 1). Some key architectural decisions include the number of cores, the type of cores (heterogenous vs homogenous), the interconnection framework, cache usage including shared cache levels vs private cache and methods of ensuring cache coherency.

These architectural decisions have a great impact on the performance as well as the resulting complexity and cost. Increasing the number of cores increases the communication framework. The communication framework uses a significant amount of space on the chip and consumes a great amount of power. The complexity of managing cache coherency between each core's private cache, the shared cache, and main memory is another challenge.

3.1 Heterogeneity vs homogeneity

As the name suggests, in Homogeneous architectures, cores are of the same type. This could result in simple, efficient, and balanced scheduling of tasks to cores as distribution to a specific core for execution is not required [17].

In heterogenous architectures, two or more unique cores are contained in the platform. The main advantage of this architecture is that tasks can be distributed to the most suitable core for execution, resulting in a higher performance. The higher performance comes with the increased complexity of task scheduling and allocation. Heterogenous architectures can also provide more efficient usage of the cores [17]. Homogenous cores may consist of multiple complex cores where each core provides higher performance at the expense of higher power consumption. Otherwise, they may also consist of multiple simple cores which use less power and provide less performance capability. A Heterogenous architecture can take advantage of both the lower power consumption of simple cores and higher performance provided by more complex cores by allocating applications to execute on the core based on their computation needs. Applications that require more performance are allocated to complex cores where applications that will not benefit as much from the complex cores can be allocated to simple cores, thereby reducing the overall power consumption of the system.

An example application that may benefit from a heterogeneous architecture is a system that must perform complex mathematical operations. A specialized core for performing these operations efficiently would improve the performance. The other tasks the system must perform would likely not require the same functionality and would benefit from a more general-purpose core. A heterogeneous platform that allows operations be scheduled to the most suitable core would be a more efficient choice.

Scheduling and allocation of tasks to the most suitable core can improve the efficiency of the execution of a program with the added cost of more complexity in the operating system. The operating system, now, must act as a match maker to schedule tasks to the core best suited for them. In addition, the scheduler must consider load balancing and avoid any idle cores. The scheduler must try to balance the loads on cores and avoid overloading any single core.

Scheduling on multicores can be divided into two categories, partitioned scheduling, and global scheduling [18]. With partition scheduling, tasks are assigned statically to the available cores. Once tasks are allocated to a core they are not allowed to switch to a different core. In global scheduling, tasks can be switched to a different core. This allows dynamically managing the distribution of tasks to the cores to ensure more even resource utilization, but switching tasks dynamically has a performance cost associated with it. The two methods can be combined in a method called semi-partitioned scheduling. In this method, tasks are initially scheduled among the cores, but allowed to be reassigned during execution to ensure cores are utilized efficiently.

3.2 Interconnection architecture

The interconnection framework facilitates the communication between the cores and the shared memory modules on the multicore chip. For higher flexibility and performance, this component is expected to have the ability to establish parallel communication paths between the cores and memory modules, and hence it has a significantly higher cost and complexity as compared to a single core processor. The cost associated with this component has a significant impact on the overall performance of the multicore processor including the size, power consumption, and latency of the memory accesses [19]. Increasing the number of cores requires more bandwidth from the interconnection architecture. The tradeoffs between the communication speed, space, and power consumption must be considered when designing the interconnection network [19].

Multicores have been designed with many different interconnection architectures. Common architectures include the bus, crossbar, and the Network on a Chip. In the bus architecture, private cache modules are all connected to common bus lines that each have a role in maintaining cache coherency [19,20]. As multiple cores can make requests over the same line at the same time, an arbiter controls access to the shared bus lines. If a core is granted access, it uses the address bus to make a request. All cores are connected to the common snoop bus, which contains the request. The impact to each core is dependent on its cache coherency protocol and may involve invalidating the core's local cache or updating its local cache. Following the snoop bus, a response bus gathers the responses from each core and broadcast them over the response bus. The responses include the impact of the request to the core (sending data on the databus, invalidate cache, etc.). Finally, the databus contains the data being read/written. The protocol is an example bus architecture and could be modified to use less lines but would require multiplexing into common bus lines [19].

In the crossbar architecture, each core is connected to each bank of shared cache [19]. A simple crossbar interconnection architecture is shown in Fig. 2 for a multicore processor with two cores and two banks of shared cache memory. With this interconnection architecture, cores can access shared memory simultaneously if the cores are accessing different banks of shared cache. In Fig. 2, each core has a unique address and data output line to the shared cache memory banks. Each shared cache memory bank has a data line which provides data to both cores. A queuing mechanism

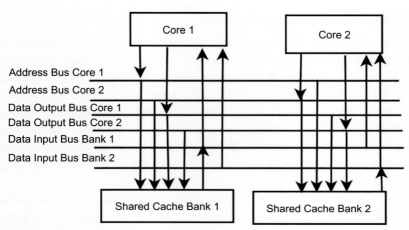

Fig. 2 Crossbar interconnection example [19].

is used to manage multiple concurrent requests from each core to the shared cache bank. This method can increase performance due to concurrent accesses to shared memory. For example, core 1 could request data from shared cache bank 2 while at the same time core 2 is writing to shared cache bank 1. As the number of cores and shared cache memory banks increase, the possible concurrent memory accesses also increase.

As multicore architectures continue to increase the number of cores, the interconnections can become the bottleneck in the system. Increasing the number of cores increases the traffic on the shared lines and increases the length of the bus lines, which in turn increases power consumption and reduced signal integrity. Point-to-point links could improve performance but would have a drastic impact on the size of the multicore chip as the number of cores increases [21]. To continue the trend of increasing the number of cores in a multicore architecture, a more scalable solution is needed.

The Network on a Chip (NoC) architecture was introduced to address the aforementioned issues [21,22]. The NoC architecture borrows concepts from computer networking to provide a more efficient and scalable alternative. Each core contains a network interface module which translates the data generated by the cores to fixed-length flow-control digits (flits) [22]. The network interface communicates the flits with a router. The routers are responsible for sending the flits in the correct direction toward the destination (as determined by the head flit). The architecture consists of several routers which route the flits to the correct adjacent router until it reaches its destination.

Fig. 3 shows a very basic NoC functional diagram. Each core contains a Network Interface module which communicates with a router [22]. The routers all communicate with one another creating a networked architecture.

The NoC uses multiple communication layers including the physical, data link, network, and transport layers [21]. The physical layer is responsible for the physical connection between resources and switches. The data link layer is responsible for the communication protocol between links on the network (between cores and switches, or between two switches). The network layer is responsible for the routing of the data over the network. The transport layer is responsible for the end-to-end communication, handling the segmentation of data on the transmitting side and reassembly of the original message on the receiving side.

With the NoC, flow control can be implemented to provide load balancing and fault handling in the network [21]. With a network, alternate routes could be chosen to reduce congestion. In addition to that benefit,

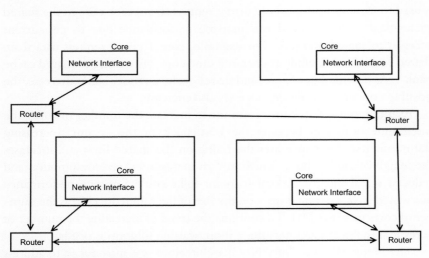

Fig. 3 NoC architecture [22].

having multiple available routes can help provide fault tolerant designs. If one communication path has failed, alternate paths could be used allowing the system to continue to work with potentially degraded performance.

Multiple switching strategies have been used to route data through an NoC. Two common methods are circuit switching and packet switching [21]. In circuit switching, a physical path is determined from the source to the destination prior to sending the data. Then all the data is sent through that same path. The advantage of the circuit switching method is data is transmitted at the same time without a delay in between sections of the data. The disadvantage is the setup time in determining the route and blocking the other traffic on the network from using those routers until the transmission is complete. If packet switching is used, data is broken into packets and each packet may take a different route. In packet switching, each packet is sent along its optimum path through the network which may be dynamic due to other traffic on the network. Unlike circuit switching, the nodes are not blocked during transmission, therefore other transmissions can share the network resources. However, an overhead is incurred because of breaking the data into packets and determining the route for each packet. The packets can be forwarded from each switch using store-and-forward, virtual cut-through, or wormhole techniques [21]. Each method has tradeoffs regarding the latency and the area consumed on the chip.

In the store-and-forward method, packets are stored in a buffer until the entire packet is received. The size of the data packet has a direct impact on the latency and the space required for storage. The virtual cut-through method does not wait for the entire packet to be stored. It forward the data to the next node as soon as the next node has space for the packet. This results in lower latency but still requires space for the storage of the entire packet.

In the wormhole method the data is divided into flits and header and trailer flits are created for each packet. The header flit is decoded by the switch to determine the path through the network, where the rest of the flits of the packet follow the header. The switches only store a small number of the flits prior to transmitting to the next router [21] and therefore do not need to store as much data at each switch. This reduces the area the switches consume compared to the store-and-forward and virtual cut-through methods.

As discussed earlier with the bus architecture, cache coherency can be maintained using methods such as snooping where each cache transaction is visible to all the cores. This method would not work using an NoC architecture [23] and becomes more difficult as the number of cores increases. A directory-based cache coherency protocol is better suited for an NoC architecture. In a directory method, status vectors can be maintained for lines of shared cache. These status vectors can indicate which cores private cache contain a copy of the shared line and if that line is modifiable in those private cores. This provides a method of maintaining cache coherency without requiring a shared bus where every core monitors each transaction.

3.3 Cache

There are many design decisions that need to be made regarding cache in a multicore architecture including the amount of cache that is shared among the cores. The cores in a multicore architecture usually have at least one level of private cache and at least one level of shared cache [24]. In some architectures there may be two levels of private cache and a one level of shared cache. The advantage of having multiple levels of private cache is the higher probability of resolving cache misses locally [25]. This advantage comes at the expense of the larger amount of memory that may be duplicated in multiple cores private cache which results in lower memory utilization and higher complexity in maintaining coherency.

Aside from determining the levels of private and shared cache, cache coherency is a large issue to consider when the system has multiple cores. It is necessary to ensure coherency between local cache memory and shared cache memory. Many different protocols have been proposed to address this issue. Managing cache coherency adds overhead in both execution time and physical resources. Two methods commonly used for managing cache coherency are bus snooping and directory [25,26].

The bus snooping method is commonly implemented on a shared bus interconnection architecture. Each core monitors the shared bus transactions and updates their private cache accordingly [26]. When a change to shared cache block is made, a controller for each core will evaluate their associated private cache and determine if it contains that shared block. If it does, an action must be taken. The action taken depends on the specific protocol used. The two common protocols are write–invalidate and write–update.

With the write-invalidate method, when a private cache block is updated, any other cache that shares that block are invalidated. The write–update method involves updating all copies of the cache block when a block is updated. Relative to the write-update method, the write-invalidate method is much more common in practice due to the reduced amount of traffic generated in performing the writes to each core's private cache [26].

Three different protocols that implement the write-invalid method are the Modified–Shared–Invalidate (MSI) states, Modified–Exclusive–Shared–Invalid (MESI), and Modified–Owned–Exclusive–Shared–Invalid (MOESI) [26]. Each protocols name indicates the transition states used. Each additional state improves performance at the expense of increased complexity.

In MSI, when a cache block is in the Shared state (S), more than one core's cache contains a copy of this cache block. When one core updates a cache block that is in the Shared state, it changes the cache block state to Modified. All other cores that have a copy of this cache block observe this change and change their states from Shared to Invalid. When a core requests to read from or write to a block of cache in the Invalid state, an action must be performed. Any core that has the cache in the Modified state must update main memory and the core requesting the cache block then obtains it from main memory.

The MESI protocol adds the Exclusive state (E). This additional state improves performance over MSI at the cost of added complexity. The Exclusive state indicates a core is the only core that has the cache block and can perform writes without the need to communicate with any other core. This reduces the overall traffic on the bus. A block of cache can be

Table 2 MOESI cache coherency states description.

State	Description
Modified (M)	Data in the local cache has been modified since being retrieved from main memory
Shared (S)	Data in the local cache has not been modified since being retrieved from main memory
Invalidated (I)	Data in the local cache is invalid. It has been modified in another core's local cache
Exclusive (E)	Indicates the local cache is the only one with the data. No other core's local cache contain this data. The data can be modified without invalidating data in another core's local cache
Owned (O)	The local cache that has the data in the owned state is the only core that can modify the data. Another core's local cache may have a copy of the data. Another local cache may get the updated data directly from the cache that has the data in the Owned state

loaded in the Exclusive state rather than the Shared state. The block in the Exclusive state can transition to the Shared state if another core later loads it, or it can move to the Modified state for a write.

The MOESI protocol adds the Owned state (O), which again improves performance at the expense of added complexity (Table 2). The Owned state indicates multiple cores have a copy of the cache block in their cache, but only the core that has the cache block in the Owned state can make modifications to it. This allows multiple cores to have a copy of a cache block. A modification to a cache block in the Owned state can be shared with other cache blocks directly without writing the modification back to main memory. This reduces the writes to main memory while still maintaining cache coherency.

MSI, MESI, and MOESI are all examples of protocols using the bus snooping method. The bus snooping method becomes less effective as the number of cores are increased given the increased amount of traffic on the shared bus. Directory based protocols are more scalable to a larger number of cores [25]. The shared cache controller maintains information regarding the usage of each cache line in each core's private cache. The controller maintains enough information to understand which core's private cache has a copy of the shared cache line and which cores have modified the shared cache line. The shared cache controller can handle coherency issues according to the implemented cache coherency protocol (write-invalidate or write-update). This method is well suited for the NoC

Table 3 Cache coherency methods.

Cache coherency method	Typical usage	Advantages/disadvantages
Bus snooping	• The cache lines are shared among the cores • Each core's local cache can monitor the shared cache line for data writes that impact the validity of the data stored in its local cache	• Suitable for a bus interconnection architecture • Bus contention issues can limit performance • Does not scale well to a larger number of cores
Directory	• Generally used in a network interconnection architecture • A cache controller monitors cache transactions • The cache controller determines when a core had valid or invalid data in its cache	• More scalable to increasing the number of cores • Avoids bus contention issues when implemented on a network • Requires a cache controller to maintain coherency

interconnection architecture given it does not require a common bus to maintain cache coherency. Table 3 compares and contrasts cache coherency methods against each other.

4. Parallel software execution

To take advantage of the capabilities of a multicore processor architecture, the software must be designed and developed for parallel execution. To achieve the maximum benefit, the processing load must be balanced on each available core in the system. Software has traditionally been written for a uniprocessor system for sequential operation. Due to the multicore architecture, methods have been introduced to exploit the potential parallelism in application programs to efficiently use each core in the system.

The approaches to software parallelism can be categorized many ways. Vandierendonck and Mens have categorized software parallelization strategies as follows [27]:

1. Explicit parallelism—Software libraries or languages are used to allow the programmers to express parallelism in the application programs. Here the programmer has control over what is executed in parallel.

2. Implicit parallelism—Parallelism is expressed using the programming language. Using traditional software languages such as C/C++, operations are specified sequentially when it may not actually be necessary for them to execute sequentially. With implicit parallelism, the programming language does not force the programmer to specify sequential operations where they are not required to be sequential. Operations that are required to be sequential are specified as sequential. Other operations which are not required to be sequential are identified for the potential to be executed in parallel.
3. Automatic parallelism—The compiler automatically extracts the parallelism from sequential code during the compilation.
4. Semiautomatic parallelization—Tools can be used to automatically identify parallelism in the application program, but the programmer is responsible for making the decision on what is done in parallel.

These four categories will be used to discuss the different approaches to implement software parallelism.

4.1 Explicit parallelism

Software can be explicitly written for parallelization using a tool that provides this capability. Open standard Application Programming Interfaces (API) exists to allow programmers to specify parallelization. One commonly used API is Open Multi-Processing (OpenMP) [28]. With OpenMP, programmers can define loops to be executed in parallel on separate cores [29]. This can be done in C code, for example, using a pragma directive. OpenMP can be used with many vendors using multiple different high-level languages such as C, C++, and Fortran [30]. The open standard API's allow the programmer to have more control over the parallelism in the executed code than is available when relying on the compiler to determine the parallelism automatically. OpenMP provides compiler directives, run-time libraries, and environmental variables [31]. All these features provide unique capabilities for the software developers to use for exploiting the parallelism available on the multicore system.

OpenMP provides the software developers the ability to specify separate threads within a program that can be executed in parallel. OpenMP also provides the ability to specify the scope of data items within the cores [31]. The data can be scoped as shared or private to a core. Using a tool like this, code can be either written for parallel execution from the start or written sequentially and later modified to provide parallelism. Naturally, writing the code with parallelism in mind from the start would be preferred.

Existing software which was originally written for serial execution could be ported to a multicore system by modifying the existing software to include the parallel operations. Explicit parallelism may be the best option for programs that are already created to operate serially and are being modified to run on a multicore processor.

4.2 Implicit parallelism

Among the four approaches, implicit parallelization would seem to be the most efficient approach. If software can be written with parallelism in mind as opposed to sequential operations, mapping the execution of different tasks to different processors would be much simpler. Starting with code written for sequential operations and explicitly determining parallelism provides opportunities to create: (i) race conditions where a data item is being modified by multiple threads which are not coordinating with each other or (ii) deadlocks where threads are waiting for the other threads indefinitely [27]. If the program is written to be executed in parallel, these situations should be avoided. One significant challenge is getting programmers to be able to adapt to a different way of thinking when writing software for parallel execution.

4.3 Automatic parallelism

Using the compiler to exploit parallelism provides several benefits. Software developers are accustomed to writing sequential software. This allows them to continue writing software sequentially while parallelism is generated by the compiler. Another benefit is the compiler can optimize the parallel execution of the software for the hardware it is going to be executed on. This allows more portable high-level software that can be used on multiple hardware platforms. The parallelism can be maximized on each hardware platform the software is running on without the need for customization. Another benefit is allowing the software developers to focus on the correctness of the software and not explicitly specifying the parallelism.

There are several techniques used by compilers for generating code optimized for the multicore architecture. Compilers are designed to optimize several different forms of parallelism including Instruction Level Parallelism, Data Level Parallelism, Loop Level Parallelism, and Pipeline Parallelism [32].

When performing a compilation, the compiler will analyze the high-level application program and detect sections of the code that can be executed in parallel. Then the compiler performs a transformation of these code sections when generating the machine code to execute in parallel.

The compiler performs this transformation with an understanding of the underlying architecture (number of cores, cache size, interconnections, etc.). The underlying architecture must be considered to transform the code for optimization based on the resources available as well as the communication costs between resources.

The compiler must perform a memory dependency analysis to exploit the Instruction Level Parallelism and Data Level Parallelism available in a program. The memory analysis must consider the communication costs between cores when optimizing parallel execution. In the case of a data dependency, it may be faster to execute instructions sequentially on one core rather than in parallel on separate cores depending on the communication costs. The compiler makes these decisions when scheduling instructions on the available cores. Memory dependency analysis look at data dependencies between iterations of loops to execute different iterations of the loop in parallel [33]. The memory dependency analysis must also look at pointers to identify possible dependencies between instructions due to pointers.

Compilers can use a technique called Decoupled Software Pipelining (DSWP) to exploit parallelism in loops with data dependencies between iterations [32]. In this method the compiler schedules instructions within iterations of the loop on different cores, such that the cores are operating in parallel on an instruction from an earlier loop iteration another core generated data for. By staggering the operations, each core is continuously busy executing instructions and the communications costs are not limiting the throughput. The communication costs do factor into the startup time of the parallel execution as the data from the early instructions must be communicated to the subsequent cores until they are all executing in parallel. This startup cost is not a significant factor when operating on a loop with many iterations.

The Decoupled Software Pipeline method is shown as compared to the Do-Across method in Fig. 4. A simple loop is shown with three instruction blocks, A, B, and C. Each instruction block is dependent on the completion of previous instruction block. Instruction block A is dependent on the previous loop iteration's instruction block A. Using a standard Do-Across method, each iteration of the loop is assigned to a core [34]. By contrast, using the Decoupled Software Pipeline approach, the instructions within iterations are assigned to different cores.

In Fig. 4, the communication costs are low and the DSWP method and Do-Across method have similar performance. There is execution time associated with obtaining the data dependencies from another core, which is being called the communication costs. Fig. 5 illustrates the advantage of

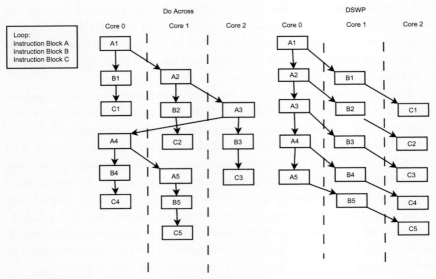

Fig. 4 DSWP vs Do-Across low communication costs.

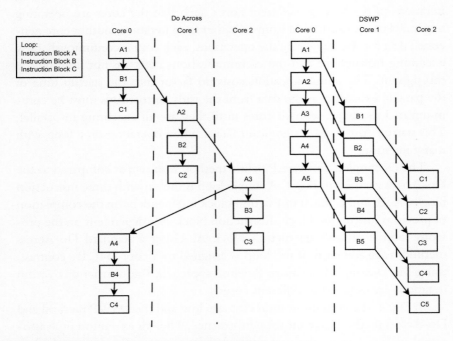

Fig. 5 DSWP vs Do-Across high communication costs.

the Decouple Software Pipeline approach when the communication costs are higher. After five iterations of the loop, it is clear the Decoupled Software Pipeline approach does a better job of minimizing the impact of the increased time to transfer dependencies among cores as compared to the Do-Across method. Five iterations of the loop are completed in the time the Do-Across method completes four iterations. This trend will continue as more iterations are performed as the Decouple Software Pipeline method reaches a state where each core is performing work and not waiting on communications from another core. In the standard Do-Across method, there is always delays waiting on the dependency from the previous loop iteration.

Parallelism can be further exploited through run-time techniques [27]. At run-time, the compiler can provide the ability to adjust using information determined at run-time. The information necessary for making these determinations can be obtained using system monitoring which gathers information about the application during execution. This information can be used to trigger events in the system to adjust to a dynamic environment where changes can be made to improve parallel execution. The execution of a program is dependent on inputs to the system as well the resource reliability. Input conditions causing a load imbalance among the cores and failures associated with one of the cores in the system are events that could trigger events at run-time to reschedule tasks among the cores. This run-time capability can improve performance beyond what is capable with only static compilation. Different compilers will behave differently [27], and the programmers still need to analyze the results of the compilations and possibly make code modifications to obtain more parallelism.

4.4 Semiautomatic parallelism

Like automatic parallelism, in semiautomatic parallelism code developed for sequential execution is analyzed to determine the available parallelism. One example where this approach would be very beneficial is updating legacy software written for sequential execution to perform parallel execution. There can be many reasons to reuse software that was originally written for sequential operation on a single core processor. A legacy design may be upgraded to use newer technology, including a multicore processor. Converting the legacy sequential software to execute in parallel could be advantages for multiple reasons including lower development costs and lower risk as the existing software has been tested and proven to operate correctly [35].

A semiautomatic technique could be used in performing the conversion from sequential software to parallel execution [35]. The software is divided into sections called "slices" that can execute in parallel on separate cores. The "slices" are determined by performing a data dependency analysis of the existing software. Code that generates intermediate data items which can be executed in parallel are identified and slices are generated from them. This provides a procedure for analyzing the code and determining the parallelism without rewriting the software and incurring the high development costs and associated risks. Although this method may not produce the maximum amount of parallelism that could be achieved by rewriting the software, it may provide sufficient parallelism to improve performance at a lower development cost (Table 4).

Table 4 Software parallelism method advantages/disadvantages.

Method	Primary advantages	Disadvantage
Explicit parallelism	• The programmers have more control over the parallelism in the software	• Significant effort required to identify parallelism • Parallelism can be identified incorrectly
Implicit parallelism	• Easier identification of parallelism in the software • Avoids unnecessary specification of sequential operations inherent in sequential programming languages	• Programmers more accustomed to programming in sequential programming languages
Automatic parallelism	• Parallelism determined by the compiler and optimized for the hardware • Software can be moved to a different hardware platform and the compiler automatically optimizes the parallelism without changes to the software	• Requires a sophisticated compiler to identify the available parallelism • It may not identify as much parallelism as other methods
Semiautomatic parallelism	• Provides an efficient method of implementing parallelism on legacy software originally written for sequential operation	• Significant effort required to perform the analysis in determining available parallelism • May not identify all true parallelism that is available

5. Conclusion

The multicore architecture has provided significant performance growth since it was first introduced. The single core processor faced significant challenges in increasing performance without significantly increasing power consumption/heat dissipation. As the size of transistors continued to shrink and operating frequencies continued to get faster, power consumption became a major issue. Multicore processors provided a method of improving performance by exploiting parallelism in the application programs. The parallel processing allowed performance improvements without continuing to increase the operating frequency and power consumption levels. This has shifted the industry toward increasing the number of cores and finding ways to exploit more parallelism in the software to improve performance.

Performance has continued to improve from multicore processors as the architectures have continued to become more complex. Large factors considered in the design of a multicore architecture include the number of cores, heterogeneity of the cores, cache size and sharing among cores, and the interconnection network used. In making each architectural decision, the cost and complexity are considered in relation to the performance benefit that can be achieved.

The increasing number of cores and the complexity of the interconnection network has led to the Network on a Chip architecture. This has been a more efficient architecture that responded to the demand of continued increases in the number of cores. The change to using a Network on a Chip has led to using the directory protocol for ensuring cache coherency among the private and shared cache memories on the chip.

To take advantage of the potential improvement the multicore architectures offer, software must exploit the available parallelism. This can be done at compile time with the compiler detecting the parallelism and incorporating that in the generated machine code. Parallelism can also be specified by the software developer using tools that are available.

The multicore architecture has had a huge impact on computer performance. It should be expected that multicore architectures will continue to provide performance improvements for many years.

References

[1] Top 500. https://top500.org, n.d. Accessed February 4, 2022.
[2] A. Roy, J. Xu, M.H. Chowdhury, Multi-core processors: a new way forward and challenges, in: 2008 International Conference on Microelectronics, Sharjah, United Arab Emirates, 2008, pp. 454–457, https://doi.org/10.1109/ICM.2008.5393510.

[3] J.D. Gelas, The quest for more processing power, part one: "is the single core CPU doomed", in: AnandTech, February 8, 2005, 2005. https://www.anandtech.com/show/1611/3. Accessed April 13, 2021.

[4] G. Blake, R.G. Dreslinski, T. Mudge, A survey of multicore processors, IEEE Signal Process. Mag. 26 (6) (2009) 26–37, https://doi.org/10.1109/MSP.2009.934110.

[5] Power4, the first multi-core, 1GHz processor, IBM's 100 Icons of Progress, August 5, 2011. https://www.ibm.com/ibm/history/ibm100/us/en/icons/power4/, n.d. Accessed August 5, 2021.

[6] AMD Athlon 64 X2 5600+ (2.8 GHz, 89W) specifications, CPU-World. https://www.cpu-world.com/CPUs/K8/AMD-Athlon%2064%20X2%205600+%20-%20ADA5600IAA6CZ%20(ADA5600CZBOX).html, n.d. Accessed October 2, 2021.

[7] Intel® Pentium® D Processor 925 [online]. https://ark.intel.com/content/www/us/en/ark/products/27517/intel-pentium-d-processor-925-4m-cache-3-00-ghz-800-mhz-fsb.html, n.d. Accessed October 2, 2021.

[8] Intel® Core™2 Quad Processor Q6600 (8M Cache, 2.4 GHz, 1066 MHz FSB). https://www.intel.com/content/www/us/en/products/sku/29765/intel-core2-quad-processor-q6600-8m-cache-2-40-ghz-1066-mhz-fsb/specifications.html, n.d. Accessed October 2, 2021.

[9] AMD Phenom II X4 965 (125W, BE) specifications, CPU-World. https://www.cpu-world.com/CPUs/K10/AMD-Phenom%20II%20X4%20965%20Black%20Edition%20-%20HDZ965FBK4DGM%20(HDZ965FBGMBOX).html, n.d. Accessed October 2, 2021.

[10] Intel® Core™i5-750 Processor. https://ark.intel.com/content/www/us/en/ark/products/42915/intel-core-i5-750-processor-8m-cache-2-66-ghz.html, n.d. Accessed October 2, 2021.

[11] AMD Phenom II X6 1090T specifications, CPU-World. https://www.cpu-world.com/CPUs/K10/AMD-Phenom%20II%20X6%201090T%20Black%20Edition%20-%20HDT90ZFBK6DGR%20(HDT90ZFBGRBOX).html, n.d. Accessed October 2, 2021.

[12] Intel® Xeon® Processor E5-2640 v3. https://ark.intel.com/content/www/us/en/ark/products/83359/intel-xeon-processor-e52640-v3-20m-cache-2-60-ghz.html, n.d. Accessed October 2, 2021.

[13] AMD Ryzen™ 7 1700X Processor. https://www.amd.com/en/products/cpu/amd-ryzen-7-1700x, n.d. Accessed October 2, 2021.

[14] Intel® Core™ i9-7900X X-series Processor. https://ark.intel.com/content/www/us/en/ark/products/123613/intel-core-i97900x-xseries-processor-13-75m-cache-up-to-4-30-ghz.html, n.d. Accessed October 2, 2021.

[15] AMD Ryzen™ Threadripper™ 1950X Processor. https://www.amd.com/en/products/cpu/amd-ryzen-threadripper-1950x, n.d. Accessed October 2, 2021.

[16] Intel® Core™ i9-7980XE Extreme Edition Processor. https://ark.intel.com/content/www/us/en/ark/products/126699/intel-core-i97980xe-extreme-edition-processor-24-75m-cache-up-to-4-20-ghz.html, n.d. Accessed October 2, 2021.

[17] D. Koufaty, D. Reddy, S. Hahn, Bias scheduling in heterogeneous multi-core architectures, in: Proc. ACM EuroSys, 2010, pp. 125–138.

[18] S. Jadon, R.S. Yadav, Multicore processor: internal structure, architecture, issues, challenges, scheduling strategies and performance, in: 2016 11th International Conference on Industrial and Information Systems (ICIIS), 2016, pp. 381–386, https://doi.org/10.1109/ICIINFS.2016.8262970.

[19] R. Kumar, V. Zyuban, D.M. Tullsen, Interconnections in multi-core architectures: understanding mechanisms, overheads and scaling, in: 32nd International Symposium on Computer Architecture (ISCA'05), Madison, WI, USA, 2005, pp. 408–419, https://doi.org/10.1109/ISCA.2005.34.

[20] S. Akram, A. Papakonstantinou, R. Kumar, D. Chen, A workload-adaptive and reconfigurable bus architecture for multicore processors, Int. J. Reconfigurable Comput. 2010 (2010) 205852, https://doi.org/10.1155/2010/205852.
[21] S. Kundu, S. Chattopadhyay, Network-on-Chip: The Next Generation of System-on-Chip Integration, first ed., CRC Press, 2015, https://doi.org/10.1201/9781315216072.
[22] W.C. Tsai, Y.C. Lan, Y.H. Hu, S.J. Chen, Networks on chips: structure and design methodologies, J. Electr. Comput. Eng. 2012 (2012) 509465, https://doi.org/10.1155/2012/509465.
[23] E. Bolotin, Z. Guz, I. Cidon, R. Ginosar, A. Kolodny, The power of priority: NoC based distributed cache coherency, in: First International Symposium on Networks-on-Chip (NOCS'07), Princeton, NJ, USA, 2007, pp. 117–126, https://doi.org/10.1109/NOCS.2007.42.
[24] A.D. Joshi, N. Ramasubramanian, Comparison of significant issues in multicore cache coherence, in: 2015 International Conference on Green Computing and Internet of Things (ICGCIoT), 2015, pp. 108–112, https://doi.org/10.1109/ICGCIoT.2015.7380439.
[25] R.E. Ahmed, M.K. Dhodhi, Directory-based cache coherence protocol for power-aware chip-multiprocessors, in: 2011 24th Canadian Conference on Electrical and Computer Engineering (CCECE), Niagara Falls, ON, Canada, 2011, pp. 001036–001039, https://doi.org/10.1109/CCECE.2011.6030618.
[26] D.P. Kaur, V. Sulochana, Design and implementation of cache coherence protocol for high-speed multiprocessor system, in: 2018 2nd IEEE International Conference on Power Electronics, Intelligent Control and Energy Systems (ICPEICES), Delhi, India, 2018, pp. 1097–1102, https://doi.org/10.1109/ICPEICES.2018.8897321.
[27] H. Vandierendonck, T. Mens, Techniques and tools for parallelizing software, IEEE Softw. 29 (2) (2012) 22–25, https://doi.org/10.1109/MS.2012.43.
[28] OpenMP, The OpenMP API Specification for Parallel Programming. https://www.openmp.org/, n.d. Accessed April 13, 2021.
[29] S.M. Alnaeli, A.D. Ali Taha, S.B. Binder, Middleware and multicore architecture: challenges and potential enhancements from software engineering perspective, in: 2016 IEEE International Conference on Electro Information Technology (EIT), Grand Forks, ND, USA, 2016, pp. 0700–0706, https://doi.org/10.1109/EIT.2016.7535325.
[30] OpenMP, OpenMP Compilers & Tools. https://www.openmp.org/resources/openmp-compilers-tools/, n.d. Accessed April 13, 2021.
[31] OpenMP Tutorial, Livermore Computing Center. https://hpc.llnl.gov/openmp-tutorial, n.d. Accessed April 13, 2021.
[32] M. Mehrara, T. Jablin, D. Upton, D. August, K. Hazelwood, S. Mahlke, Multicore compilation strategies and challenges, IEEE Signal Process. Mag. 26 (6) (2009) 55–63, https://doi.org/10.1109/MSP.2009.934117.
[33] J.T. Lim, A.R. Hurson, K.M. Kavi, B. Lee, A loop allocation policy for DOACROSS loops, in: Symposium in Parallel and Distributed Processing, 1996, pp. 240–249.
[34] G. Ottoni, R. Rangan, A. Stoler, D.I. August, Automatic thread extraction with decoupled software pipelining, in: 38th Annual IEEE/ACM International Symposium on Microarchitecture (MICRO'05), 2005, pp. 105–118, https://doi.org/10.1109/MICRO.2005.13.
[35] T.R. Vinay, A.A. Chikkamannur, A methodology for migration of software from single-core to multi-core machine, in: 2016 International Conference on Computation System and Information Technology for Sustainable Solutions (CSITSS), Bengaluru, India, 2016, pp. 367–369, https://doi.org/10.1109/CSITSS.2016.7779388.

About the authors

Tim Guertin received a M.S. in Computer Engineering from Missouri University of Science and Technology in 2021 and a B.S. in Electrical Engineering from St. Cloud State University in 2007. He has more than 15 years of experience as a Software and Systems Engineer in the aerospace industry. His research interests include computer architecture, parallel computing, and machine learning.

Ali Hurson is an Electrical and Computer Engineering Professor at Missouri University of Science and Technology. His research in the past has been supported by various government agencies and private industries. He has more than 350 publications in the areas of databases, computer architecture, and pervasive computing.

CHAPTER FIVE

Perceptual image hashing using rotation invariant uniform local binary patterns and color feature

Ming Xia[a], Siwei Li[a], Weibing Chen[b], and Gaobo Yang[a]
[a]School of Computer Science and Electrical Engineering, Hunan University, Changsha, China
[b]College of Electronic Communication & Electrical Engineering, Changsha University, Changsha, China

Contents

Advances in Computers, Volume 130
ISSN 0065-2458
https://doi.org/10.1016/bs.adcom.2022.12.001

Abstract

Perceptual image hashing is widely used in cloud-based multimedia system for various security purposes, including image integrity authentication and tamper detection. In this chapter, we propose a robust perceptual image hashing scheme by exploiting rotation invariant uniform local binary patterns ($LBP_{P,R}^{riu2}$) and color features. In our scheme, input image is first pre-processed to build a normalized image. Next, $LBP_{P,R}^{riu2}$ is exploited to build a stable texture feature matrix, and Non-negative Matrix Factorization (NMF) is used for data reduction and for constructing the $LBP_{P,R}^{riu2}$-NMF features. Then, color features are generated from the mean value of each color block. Finally, the $LBP_{P,R}^{riu2}$-NMF features and the color features are concatenated and quantized to form final image hash, which is referred to the $LBP_{P,R}^{riu2}$-NMF-color hashing. Experimental results show that the proposed image hashing scheme has superior robustness and better discrimination capability than most existing methods.

1. Introduction

With the rapid advancement of image processing and network technologies, there are exponentially growing amount of digital image data conveyed, broadcasted and browsed over the Internet. For example, the cloud-based multimedia systems, which are attracting more and more users, are developing rapidly due to the availability of high-end computation infrastructure. However, transmitting image data to cloud servers also poses a lot of security and privacy threats [1]. In particular, digital image might be easily modified by ordinary users with more and more powerful image editing tools. Digital image forgeries and unauthorized uses have reached a significant level that makes image content security issues be very challenging and demanding. Common malicious image forgeries include object inserting, object removal, image splicing, and so on [2]. The capability to detect any image content changes or image forgeries has been very important for many applications, especially for journalistic photography, scientific report or artwork image databases [3]. The image content security issues, which demand us to check the safety of exchanged image data confidentiality, authenticity and integrity, have drawn wide concerns and attentions from the communities of both academic research and multimedia industry.

To address these security challenges, a lot of image content security techniques have been developed to prevent illegal access and unauthorized distributions of image contents. These existing techniques can be categorized into three groups: image watermarking-based schemes [4], image forensics-based schemes [2] and perceptual image hashing-based schemes [5].

- *Image watermarking schemes* have been widely used for protecting intellectual property rights. They embed visible or imperceptible signal, which is referred as watermark, into image data to form the watermarked image. Since the embedded watermarks usually carry copyright information, they should be robust against malicious attacks, so that they can be correctly extracted or 50restored to claim the ownership of the host image data. A fragile or semi-fragile watermarking scheme can detect various changes of the host image data, which provides some form of guarantee that the image data has not been tampered. A fragile watermarking scheme should be capable to identify which portions of the watermarked image data are authentic [6], and which portions are corrupted or maliciously tampered [4]. Though image watermarking can be used in copyright protection or content authentication for an individual image, it is not suitable for some applications in which a large scale image searching is required. Moreover, watermark embedding usually causes some slight distortions to the host image data.
- *Image forensics-based schemes* expose potential tampering traces in digital images without the aid of any auxiliary data, such as watermark [2]. Since the traces for various/image forgeries are usually unnoticeable, the core of passive image forensics is to find out some appropriate forensics clues, from which some statistical features are specifically designed and then extracted for the binary classification of image authenticity [2,3]. In general, the forensics features are designed to be closely related with the specific type of possible image forgery operation. That is, image forensic techniques typically depend on the similarity/inconsistency of the intrinsic physical features and/or the inherent signal statistics, which are used to expose potential image forgery operations [7]. However, the lack of any auxiliary data/side information about the original image data makes image forensics be computation-intensive and experience-dependent. Moreover, since the designed forensics features are usually specific to the type of possible image forgeries, it is still quite difficult to achieve universal or general-purpose forensics for multiple image forgeries.
- *Image hashing-based schemes* construct a hash value based on the visual content of an input image [5]. A cryptographic hash function is a special class of hash function that meets certain properties, which makes it be suitable for the use in cryptography [8]. In essence, the cryptographic hash function is a mathematical algorithm which maps the image data of arbitrary size into a bit string of a fixed size (a hash), which is

specifically designed to be a one-way irreversible function. The traditional cryptographic hash functions including MD5 and SHA-1 are sensitive to even one bit change, which make them be suitable for data integrity and retrieval [9]. The traditional cryptosystems might be exploited to encrypt image data, but it is still not an ideal choice for the following two reasons. First, the size of the image data is usually large, and thus the traditional cryptosystems need quite much time to encrypt the image data. Second, the decrypted image data must be equal to the original image data. However, this requirement is often not necessary for the image data. As an alternative, perceptual image hashing produces a snippet or fingerprint of various image contents [10]. That is, the purpose of the perceptual image hashing is totally different from the traditional cryptographic hash function. It is well-known that digital images might undergo content-preserving manipulations such as image compression, cropping, contrast enhancement and scaling/resizing. Perceptual image hashing should tolerate such types of acceptable content-preserving changes. For those images with almost the same visual appearances, similar hash values should be produced for them. Meanwhile, perceptual image hashing should also be sensitive to malicious content-changing manipulations and attacks such as removing semantic objects from input image or adding some new objects into it, changing image background or lighting conditions. Therefore, a perceptual image hashing scheme must compromise well between perceptual robustness and discriminative capability. Up to present, perceptual image hashing has been widely used in many image applications, which include image authentication [5], content-based image retrieval, image indexing [11], copy/near duplicate detection [12], image forgery detection [13], and even image quality assessment [14].

In general, there are two key steps, which are illustrated in Fig. 1, for a perceptual image hashing scheme. The first step is feature extraction, which

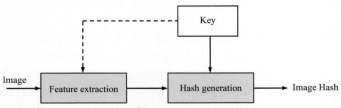

Fig. 1 Two key steps of the image hashing scheme. *From Davarzani, R., Mozaffari, S., Yaghmaie, K. Perceptual image hashing using center-symmetric local binary patterns. Multimed. Tools Appl. 2016, 75, (8), pp. 4639–4667.*

extracts a feature vector from the input image. The extracted features usually depend on input image content. The second step is hash generation, in which the feature vector is compressed and quantized into a binary or real number sequence to form the final hash value. Since the image hash might serve as a secure tag, a secret key can be incorporated into the step of either feature extraction or hash generation, which guarantees that the final hash value is difficult to be obtained by unauthorized adversaries without knowing the secret key in advance. Some works argue that a perceptual image hashing system generally consists of four pipeline stages: the transformation stage, the feature extraction stage, the quantization stage and the compression and encryption stage [8]. Actually, the transformation stage can be regarded as some kind of image pre-processing for feature extraction, since the input image undergoes special and/or frequency transformation to make the extracted features depend the values of image pixels or the frequency domain coefficients. In the compression and encryption stage, the binary hash string is compressed and encrypted into the final perceptual hash which is much shorter. Apparently, this can be regarded as the essential step for hash generation. Therefore, all perceptual image hashing systems share the same key techniques, no matter two key steps or four pipeline stages.

In this chapter, we investigate the use of rotation invariant uniform local binary patterns ($LBP_{P,R}^{riu2}$) for perceptual image hashing. Specifically, $LBP_{P,R}^{riu2}$ is exploited in the step of feature extraction to build a stable texture feature matrix. Then, Non-negative Matrix Factorization (NMF) is exploited for data reduction. Thus, the $LBP_{P,R}^{riu2}$-NMF features are constructed. Moreover, some color features are generated from the mean value of each color block. The $LBP_{P,R}^{riu2}$-NMF features and the color feature are then concatenated to form the feature vector, which is referred to the $LBP_{P,R}^{riu2}$-NMF-Color feature vector. In the hash generation step, the extracted $LBP_{P,R}^{riu2}$-NMF-Color feature vector is quantized into a binary sequence as the final hash. To alleviate the influences of content-preserving image manipulations, image pre-processing operations, which include image normalization and Gaussian low-pass filtering, are conducted before feature extraction. Note that the final hash has fixed length.

The rest of this chapter is organized as follows: Section 2 summaries some basic concepts of perceptual image hashing. Section 3 surveys the existing perceptual image hashing works. In Section 4, rotation invariant uniform local binary patterns ($LBP_{P,R}^{riu2}$) and NMF are briefly introduced. Section 5 presents the proposed perceptual image hashing scheme. Section 5 reports the experimental results and analysis. Section 6 concludes our research.

2. Basic concepts of perceptual image hashing

2.1 Content-preserving and content-changing manipulations for perceptual image hashing

Perceptual Image hashing extracts certain features from an input image, and then calculates a hash value based on the extracted features. By comparing the hash value of the original image and the hash value of the image to be authenticated, the integrity and authentication of the image content can be well verified. The terminology of "perceptual hash" means that the perceived quality of image content after permissible image tampering should be perceptually acceptable. That is, perceptual hash functions are expected to be able to survive after acceptable content–preserving image manipulations, whereas reject malicious image manipulations such as those image forgeries that change semantic image contents. In Table 1, we briefly summarize the most representative content–preserving and content–changing image manipulations for perceptible image hashing [8]. Note that the content–preserving and content–changing image manipulations defined here are very similar with those for semi-fragile image watermarking, in which fragile watermarks are designed to be easily destroyed for any image manipulation even when the watermarked image is manipulated in the slightest manner, and robust watermarks are designed to be capable of tolerating some degree of changes to the watermarked image. That is, robust watermarks are designed

Table 1 A rough classification of content-preserving and content-changing image manipulations.

Content-preserving manipulations	Content-changing manipulations
– Transmission errors – Noise addition – JPEG Compression and quantization – Sampling – Rotation – Retargetting including scaling and cropping – γDistortion – Changes of brightness, hue and saturation – Contrast adjustment	– Removing object from image – Adding new objects – Changing the position of motion object – Changes of image characteristics such as color, textures and structure, etc. – Changes of image background: day time or location – Changes of light conditions: shadow manipulations, etc.

From Davarzani, R., Mozaffari, S., Yaghmaie, K. Perceptual image hashing using center-symmetric local binary patterns. Multimed. Tools Appl. 2016, 75, (8), pp. 4639–4667 to be able to survive on content-preserving manipulations. However, for passive image forensics, both content-preserving and content-changing image manipulations are usually regarded as image forgeries. That is, even those content-preserving image manipulations such as double JPEG compression [15], contrast enhancement [16] and image retargetting/resizing [17,18] are investigated so as to develop specific detection tools for them, respectively. Nevertheless, if powerful tools or advanced techniques are developed to expose those content-changing image manipulations such as copy-move forgery [19] and object removal by image inpainting [2], passive image forensics will be much more meaningful for practical use.

2.2 Metrics and important requirements of perceptual image hashing

Let I be the input image. Its perceptually similar version and content-different version are defined as I_{ident} and I_{diff}, respectively. Let $P\,00000$ be the probability and H_k be an image hash function depending on a secret key k. Assume that the hash function H_k produces a binary hash string of length l from the input image. For a perceptual image hashing scheme, it should meet some desirable properties, which are summarized as follows [20]:

(1) Perceptual Robustness.
It means that perceptually identical images should have similar hashes. That is,

$$P(H_k(I) \approx H_k(I_{ident})) \geq 1 - \varepsilon, \quad 0 \leq \varepsilon < 1 \tag{1}$$

(2) Uniqueness
It implies that perceptually distinct images should have unique signatures. That is,

$$P(H_k(I) \neq H_k(I_{ident})) \geq 1 - \theta, \quad 0 \leq 0 < 1, \tag{2}$$

(3) Unpredictability
It means equal distribution of hash values. That is,

$$P(H_k(I) = h_I) \approx \frac{1}{2^l}, \quad \forall h_I \in \{0, 1\}^l \tag{3}$$

where h_I is the l-bit binary hash value for the image I.

(4) Compactness
It means that the size of the hash signature should be much smaller than the original image I. That is,

$$Size(H_k(I)) \ll Size(I) \tag{4}$$

(5) Pair-wise independence for perceptually different images I and I_{doff}

$$P\big(H_k(I) = H_I | H_k\big(I_{diff}\big) = h_{I_{diff}}\big) \approx P(H_k(I_{ident}) = h_I), \quad \forall h_I, h_{I_{diff}} \in \{0, 1\}^l$$

$$(5)$$

Please note that the above five requirements are usually conflicting. Thus, the design of efficient yet robust perceptual image hashing techniques is a challenging problem since it should address the compromise among various conflicting requirements. For example, to meet property in Eq. ($_1$), most perceptual hash functions try to extract features of images which are invariant under insignificant global modifications such as image compression or enhancement. Eq. (2) means that, given an image I, it should be nearly impossible for an adversary to construct a perceptually different image I_{diff} such that $H(I) = H(I_{diff})$. This property can be hard to achieve because the features used by published perceptual hash functions are publicly known. Likewise, for perfect unpredictability, an equal distribution (Eq. $_3$) of the hash values is needed. This would deter achieving the property in Eq. ($_1$). Therefore, perceptual hash functions have to achieve these conflicting properties to some extent and/or facilitate trade-offs, which depends on the application. From a practical point of view, both robustness and security are important for perceptual image hashing.

3. Literature review and research gaps

In recent years, there has been a growing number of research works on perceptual image hashing. As claimed earlier, there are two key steps for any perceptual image hashing scheme. The first step is feature extraction, which provides a compact yet robust representation of image content. The second step is hash generation, which encodes the extracted feature vector into a binary string or a real number sequence to form the final hash value. According to the strategy for feature extraction, existing perceptual image hashing approaches can be roughly divided into three categories: transform based schemes, dimension reduction based schemes, and local feature pattern based schemes [20].

3.1 Transform based schemes

Image transform is a common image processing operation, which is exploited to convert an input image from one domain to another. Classical image transformations such as Discrete Cosine Transform (DCT), Discrete Fourier Transform (DFT) and Discrete Wavelet Transform (DWT) have been widely used for feature extraction in various pattern classification/recognition tasks. Actually, the excellent properties of DCT can be used to generate perceptual image hash function. For example, it is well-known that the low-frequency DCT coefficients of an image are quite stable under various image manipulations. For two different 8×8 blocks within an image, their DCT coefficients at the same position can represent their invariant relationships before and after JPEG compression. By exploiting the stability of the low-frequency DCT coefficients, Fridrich et al. proposed a robust image hashing algorithm to verify the authentication and integrity of still images [21]. By exploiting the invariance of the relationships between DCT coefficients at the same position in separate blocks of an image, Lin and Chang proposed a robust image hashing scheme, which can distinguish JPEG compression from malicious manipulations [22]. However, this image hashing is still vulnerable to some other perceptually insignificant image manipulations such as blurring. Zhang Bin et al. proposed an anti-JPEG compression image perceptual hashing algorithm [23]. It exploits the fact that digital images are JPEG compressed mainly by quantifying the DCT coefficients, which removes the higher frequency DCT coefficients while protecting the lower frequency values. Specifically, the relative stability of the low frequency DCT coefficients in JPEG compression is exploited for perceptual image hashing. It can not only achieve the authentication of JPEG image, and also can effectively detect some malicious manipulations including image tampering and noise. Yu and Sun also proposed a robust image hashing based on the signs of DCT coefficients [24]. From an image with reduced dimensions, 2-D DCT is applied to derive an initial feature vector. Then, the sign bit of this feature vector is extracted to form an intermediate hash, which is incorporated with some security mechanism to derive the final hash. In addition, Venkatesan et al. proposed a perceptual image hashing scheme by exploiting the quantized statistics of randomized rectangles in DWT domain [25]. The averages or variances of the random blocks in wavelet image are computed and then quantized using a randomized rounding to form a secure binary hash. It is robust against some geometric attacks, but is still sensitive to some other image manipulations such as contrast adjustment and gamma correction.

The Fourier-Mellin Transform (FMT) is a translation, scale and rotation invariant transform, which is also widely used in pattern recognition. Swaminathan et al. proposed a robust and secure image hashing [26], in which the final image hash is generated based on Fourier transform features

and controlled randomization. The robustness of image hashing is formu-
lated as a hypothesis testing problem. This method is resilient to various
content-preserving manipulations such as geometric distortions, filtering
operations, and etc. A general framework is also proposed to evaluate the
security issues of image hashing schemes, in which the hash values are
modeled as random variables, and the uncertainty is quantified in terms of
the differential entropy of hash values. Later, another perceptual image
hashing approach was also proposed based on FMT [27]. Its basic idea is
to exploit the invariance property of the FMT, in which robust hashes
are extracted by using overlapping blocks in the transform domain. To
secure the system, two secrete keys are used to randomly select the blocks
and the low-frequency coefficients within these blocks. It proves desirable
robustness against signal processing operations and geometric attacks including rotation, scale and translation (RST). It also achieves better robustness
against geometric distortions than other transform based image hashing techniques. Another secure and robust image hashing method was proposed
based on the FMT and compressive sensing [28]. Specifically, FMT is used
to provide robustness against RST attacks, and the property of dimension
reduction inherent in compressive sensing is exploited for hash design.
This image hashing method is computationally secure without any additional randomization process.

In recent years, Radon transform has also been used for image hashing
[29–31]. Radon transform is used to divide the image into radial
projections. A set of radial projections of the image pixels compute the
RAdial Variance (RAV) vector. Then, the discrete cosine transform of
the RAV vector produce the Transformed RAV (TRAV) vector, whose
first 40 coefficients define our robust image feature vector, denoted as
Radial hASH (RASH) vector [30]. This image hashing method is robust
against scaling and rotation, but its discriminative capability still needs to
be improved. To exploit the advantages of both the frequency localization
properties of DWT and the shift/rotation invariant property of the Radon
transform, Guo et al. proposed a content based image hashing via wavelet
and Radon transform [31]. Robust features are first extracted from the
DWT coefficients followed by Radon transform, and then a probabilistic
quantization is used to map the feature values into a binary sequence. It
can not only resist perceptually insignificant modifications such as JPEG
compression, filtering, scaling and rotation, but also detect content-
changing attacks such as object inserting. Later, Lei and Wang proposed a
robust image hashing scheme in Radon transform domain for image

authentication [29]. First, it performs Radon transform on the image, and calculates the moment features which are invariant to translation and scaling in the projection space. Then, DFT is applied on the moments features so as to resist image rotation attack. Finally, the magnitude of the significant DFT coefficients is normalized and quantized as the image hash bits. It is robust to typical image manipulations including compression, geometric distortion, blur, enhancement and noise.

3.2 Dimension-reduction based schemes

Dimension reduction refers to the process of converting a set of data having vast dimensions into data with much lesser dimensions ensuring that it conveys similar information in a concise way. If the data set to be reduced contains statistical image features, dimension reduction is also referred to be feature transformation, which reduces the dimensionality in the feature data by transforming them into new feature data. For image hashing scheme, various dimension reduction strategies have been used to save the storage space for image hash. In this category, Singular Value Decomposition (SVD) and Non-negative Matrix Factorization (NMF) are two popular techniques. In linear algebra, SVD is originally a factorization of a real or complex matrix. Thus, SVD is also called as SVD decomposition. For image hashing, SVD is usually exploited to generate the intermediate hash value. NMF is another dimension reduction technique, which is based on the low-rank approximation of the feature space. Thus, NMF is also called as non-negative matrix approximation. For this category of image hashing schemes, the advantages of matrix factorization or decomposition, i.e., SVD or NMF, are used to extract image content-based features. The most representative works are summarized as follows.

By considering images (as well as attacks on them) as a sequence of linear operators, Kozat et al. proposed a robust perceptual image hashing via matrix invariants [32]. First, a secondary image is constructed from the input image by pseudo-randomly extracting features which approximately capture semi-global geometric characteristics. Then, the final features are further extracted from the secondary image (which does not perceptually resemble the input), which can be further suitably quantized as a hash value. Specifically, spectral matrix invariants are used, which are embodied by SVD. Due to the well-captured geometric structure of images, this hashing method is robust against some severe geometric disturbances such as rotation, scaling and cropping. Later, Lahouari proposed a robust perceptual hashing scheme for color image [33]. It treats the image color components in a holistic manner by exploiting the intrinsic color correlation between the image components. First, a compact representation of color images is provided by handling the red, green and blue (RGB) components as a single entity using hyper-complex representations. Then, the ability of Quaternion SVD (Q-SVD) provides the best low-rank approximation of quaternion matrices in the sense of Frobenius norm. Finally, possible geometric attacks are properly modeled as an independent and identically-distributed hyper-complex noise on the singular vectors. This method outperforms most existing SVD-based image hashing schemes in terms of lower miss and false alarm probabilities.

In 2007, a secure and robust image hashing scheme is proposed by exploiting NMF [34]. An input image is regarded as matrices, and the goal of image hashing is to find a randomized dimensionality reduction that retains the essence of the original image matrix while preventing intentional attacks of guessing and forgery. Motivated by the fact that standard-rank reduction techniques such as SVD produce low-rank bases without considering the structure (i.e., non-negativity for images) of the original data, two desirable properties of NMFs are exploited for secure image hashing. First, the additivity property resulting from the non-negativity constraints results in bases that capture local characteristics of the image, thereby significantly reducing mis-classification. Second, the effect of geometric attacks on images in the spatial domain manifests (approximately) as independent identically distributed noise on NMF vectors, allowing the design of detectors that are both computationally simple and at the same time optimal in the sense of minimizing error probabilities. The proposed image hashing scheme can tolerate many perceptually insignificant attacks, but it is still vulnerable to brightness changes and large geometric transformations. Later, Tang et al. proposed a lexicographical-structured framework to generate image hashes [35]. There are mainly two components for image hashing: dictionary construction and maintenance, and hash generation. The dictionary is a large collection of feature vectors called words, representing the characteristics of various image blocks, which is used to provide basic building blocks to form the hash. For the hash generation, the blocks of the input image are represented by those features associated to the sub-dictionaries. Under the framework, DCT and NMF are used to implement the hashing scheme. Besides, fast Johnson-Lindenstrauss transform (FJLT), which can be categorized into the dimension reduction group, is used for image hashing. This image hashing scheme is resistant to normal content-preserving manipulations, and has a very low collision probability. Furthermore, Tang et al. proposed a robust perceptual image hashing based on ring partition and NMF [36]. Its key contribution is the construction of rotation-invariant secondary image by ring partition, which makes image hash be resistant to rotation. In addition, the NMF coefficients are approximately linearly changed by some content-preserving image manipulations. The hash similarity is measured by correlation coefficient. It claims that this image hashing scheme is robust against content-preserving operations such as rotation, JPEG compression, Gaussian low-pass filtering, brightness adjustment, gamma correction and image scaling.

Security is one of the crucial issues for perceptual image hashing. Due to the robustness requirement, the key disclosure issue is also closely associated with all perceptual image hashing techniques. For the security of the NMF-based perceptual image hashing, a preliminary investigation was made in 2010 [37]. It claims that although three independent keys can be used in the different stages of the NMF-based perceptual image hashing scheme, the first key plays a crucial role to ensure the hashing system. Furthermore, an efficient technique has been proposed to estimate the first key based on the observation of image/hash pairs. Specifically, it has been shown that when the first key is reused several times on images with different visual content, it might be accurately estimated. However, the use of a secret key combined with image-dependent keys can enhance security in the sense that the information leakage about the final key will be smaller due to its dependence on the image content.

3.3 Local feature pattern based schemes

Image features are often divided into two categories: local features and global features. Some existing image hashing methods for image searching and retrieval are based on global feature representations, which are susceptible to image variations such as viewpoint changes and background cluttering [38]. Moreover, traditional global representations gather local features directly to output a single vector without the analysis of the intrinsic geometric property of local features. That is, the image hashing techniques that use only global features have only limited discriminative capability. Local feature patterns have been widely investigated in various computer vision applications such as robust matching. Local feature patterns, which represent the local content of each image, can also be exploited for perceptual image hashing. To the best of our knowledge, the widely-used local features in image hashing include edges, corners, blobs, salient regions and key points. The main advantages of using local features in image hashing are their robustness against geometric distortions, including rotation, scaling, and shearing. Yan et al. presented a multiscale image hashing scheme for robust tampering detection [13]. It exploits the location-context information, which are obtained by adaptive and local feature extraction techniques. Specifically, the global and color hash component is used to determine whether the received image has the same contents as the trusted one or has been maliciously tampered, or just visually different. If the received image is judged as being tampered, the tampered regions are localized by using the multiscale hash component. However, this hashing scheme is only robust against content-preserving attacks, including both common signal processing and geometric distortions. Another shape-contexts-based image hashing approach was proposed by exploiting robust local feature points [39]. Specifically, the robust SIFT-Harris detector is exploited to select the most stable SIFT key points under various content-preserving distortions, and then compact and robust image hashes are generated by embedding the detected local features into shape-contexts-based descriptors. Because of the robust salient key points detection and the shape-contexts--based feature descriptors, the proposed image hashing scheme is robust to a wide range of distortions and attacks, and can also be used for image tampering detection. In 2016, an unsupervised bilinear local hashing (UBLH) was proposed for image similarity search [38]. Instead of using a single large projection matrix, local features are projected from a high-dimensional space to a lower-dimensional Hamming space via compact bilinear projections. UBLH takes the matrix expression of local features as input and preserves well the feature-to-feature and image-to-image structures of local features simultaneously. Recently, Davarzani et al. presented a perceptual image hashing scheme by exploiting center-symmetric local binary patterns (CSLBP) [20]. Specifically, the CSLBP features are extracted from each non-overlapping block within the original gray-scale image. For each block, the final hash code is obtained by inner product of its CSLBP feature vector and a pseudo-random weight vector. Furthermore, SVD is combined with CSLBP for hash generation. Thus, this image hashing scheme is also called SVD-CSLBP. It achieves desirable perceptual robustness against a lot of attacks such as additive noise, blurring, brightness changes and JPEG compression. However, its discrimination capability can be further improved.

However, local features have their own disadvantages in image hashing. First, local features are sensitive to classical attacks such as additive noise, blurring and compression, which might make them be unsuitable for practical image hashing. Second, local feature extraction is a time-consuming task. Thus, how to improve the strength of local feature patterns against the above-mentioned attacks is still an open issue in the community of perceptual image hashing. Actually, the combination of global and local features is promising for perceptual image hashing. In 2016, a perceptually robust image hashing scheme was proposed based on both global and local features [40]. Global features are extracted by using ring partition, which are then projected by gradient non–negative matrix factorization (PGNMF). The ring partitioning technique converts a square image into a secondary image which is rotation invariant. Meanwhile, PGNMF is usually much faster than other NMFs for dimension reduction. Local image features are extracted from salient regions, which consist of both position and texture information. Experimental results show that the combination of global and local features makes the image hashing scheme be robust against acceptable content-preserving image operations. Moreover, PGNMF enhances the discriminative capability.

4. Our image hashing approach

In this sector, we propose a novel image hashing approach by exploiting both rotation invariant uniform local binary patterns ($LBP_{P,\,R}^{riu2}$) and color feature. Fig. 2 is the flowchart of our approach. The proposed image hashing scheme includes four key steps, i.e., pre-processing, $LBP_{P,\,R}^{riu2}$ feature extraction, NMF and color feature extraction. Compared with existing image hashing works, the contributions of the proposed approach is different with design of twofolds. First, $LBP_{P,\,R}^{riu2}$ is exploited to construct a stable texture feature matrix, which makes image hash representing well image content. NMF supports compact representation of feature matrix,

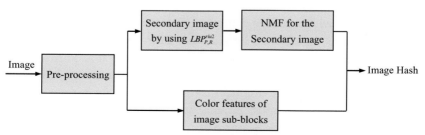

Fig. 2 Flowchart of the proposed image hashing approach. *Source: author.*

which reduces the length of final image hash. Second, since the color feature by averaging the R,G,B channels is combined with $LBP_{P,\ R}^{riu2}$ to construct image hash, which improves the discrimination capability. The details of these steps are presented in Fig. 2.

4.1 Pre-processing

To alleviate the side effects of permissible content-preserving manipulations, image pre-processing is conducted here. First, bilinear interpolation is used to resize the input image into a standard size of M × M. This makes image hashing be resistant to image resizing, and the final hash will have fixed length. Then, the color image is preserved and then converted into a gray-scale image I_0 simultaneously. The reason behind the fact that we preserve the color images I is that color information is needed for successive color feature extraction. Finally, a Gaussian low-pass filtering is applied to the resized image I_0 to obtain the blurred image I_1, which mitigates the influences of noise and filtering on input image.

4.2 $LBP_{P,\ R}^{riu2}$

Local Binary Pattern (LBP) is a simple yet efficient local texture descriptor, which has been widely used in image analysis due to its superior performance. Given a pixel in the image, an LBP code is computed by comparing it with its neighboring pixels:

$$LBP_{P,R} = \sum_{p=0}^{p-1} s\left(g_p - g_c\right) 2^p, \quad s(x) = \begin{cases} 1 & x \geq 0 \\ o & x < 0 \end{cases} \tag{6}$$

where g_c is the grayscale value of the central pixel, g_p is the grayscale value of its neighbors, P is the total number of neighboring pixels involved in the computation, and R is the circle radius of neighborhood. Usually, the sample number $P=8$ is the most commonly used, with the circle radius $R=1$. Thus, current pixel is encoded into an 8-bit integer value. By sweeping over the whole image with recursive computation of the LBP value of each pixel, the input image is transformed into another LBP image.

Though LBP is an excellent local texture descriptor, it is quite sensitive to rotation. Later, numerous modifications and improvements have been suggested on the basis of the original LBP methodology for various applications. Among them, the most popular improvements are uniform local binary patterns and rotation invariant uniform local binary patterns.

Please note that the term "uniform" refers to the uniform appearance of the local binary pattern. That is, there is only limited number of transitions or discontinuities in the circular presentation of the pattern. Especially, uniform local binary patterns are patterns with at most two circular 0–1 and 1–0 transitions. For example, patterns 001110000, 11111111, 00000000, and 11011111 are uniform, whereas patterns 01010000, 01001110 or 10101100 are not uniform. Selecting only uniform patterns contributes to both reducing the length of the feature vector (LBP histogram) and improving the classification performance when using the LBP features. In 2002, Ojala presented a new methodology of grayscale and rotation invariant uniform local binary patterns, which is referred to be $LBP_{P,R}^{riu2}$. This operator is derived from the joint distribution of gray values of a circularly symmetric neighbor set of pixels in a local neighborhood, and thus it is invariant against any monotonic transformation of the gray scale. Rotation invariance is achieved by recognizing that this gray scale invariant operator incorporates a fixed set of rotation invariant patterns.

From Eq.(6), the $LBP_{P,R}$ operator produces 2^P output values, corresponding to the 2^P different binary patterns that can be formed by the P pixels in the neighbor set. When the image is rotated, the gray values g_p will correspondingly move along the perimeter of the circle around g_0. Since g_0 is always assigned to be the gray value of element $(0, R)$, to the right of g_c, rotating a particular binary pattern naturally results in a different $LBP_{P,R}$ value. To remove the effect of rotation, rotation invariant uniform local binary patterns are first defined on the basis of local binary patterns as follows.

$$LBP_{P,R}^{ri} = \min\{ROR(LBP_{R,R}, i)|i = 0, 1, ..., P-1\} \qquad (7)$$

where the superscript ri stands for rotation invariant, P is the number of involved neighboring pixels, and R is the radius of neighborhood of adjacent pixels. $ROR(x, i)$ represents a circular bit-wise right shift on the P-bit number xi times. In terms of image pixels, Eq. (6) simply corresponds to rotating the neighbor set clockwise so many times that a maximal number of the most significant bits, starting from g_{P-1}, are 0. Since $LBP_{P,R}^{ri}$ quantifies the occurrence statistics of individual rotation invariant patterns, which correspond to certain micro features in the image, the $LBP_{P,R}^{ri}$ patterns can be regarded as feature detectors.

Since the signed differences g_p-g_c are not affected by changes in mean luminance, the joint difference distribution is invariant against grayscale shifts. To further achieve grayscale invariance, the differences of g_p-g_c are just considered with their signs of the differences, instead of their exact values. To formally define the "uniform" patterns, a uniformity measure U(pattern) is introduced, which corresponds to the number of spatial transitions (bit-wise 0/1 changes) in the "pattern." That is, uniform local binary pattern is defined as

$$U(LBP_{P,R}) = |s(g_{P-1} - g_c) - s(g_0 - g_c)|$$

$$+ \sum_{p=1}^{P-1} |s(g_p - g_c) - s(g_{p-1} - g_c)| \tag{8}$$

$$s(x) = \begin{cases} 1 & x \geq 0 \\ 0 & x < 0 \end{cases} \tag{9}$$

where g_c is the gray value of the central pixel, g_i is the value of its neighbors. The value of U is actually the number of spatial transitions. Then, the rotation invariant uniform local binary patterns, which are referred to be $LBP_{P,R}^{riu2}$, is defined as follows.

$$LBP_{P,R}^{riu2} = \begin{cases} \sum_{i=0}^{P-1} s(g_i - g_c), & if\ U(LBP_{P,R} \leq 2) \\ P+1 & otherwise \end{cases} \tag{10}$$

The superscript $riu2$ reflects the use of rotation invariant "uniform" patterns that have U value of at most 2. By definition, exactly $P+1$ "uniform" binary patterns can occur in a circularly symmetric neighbor set of P pixels. Eq. (10) assigns a unique label to each of them, corresponding to the number of "1" bits in the pattern, while the "nonuniform" patterns are grouped under the "miscellaneous" label $(P+1)$. In practice, the mapping from $LBP_{P,R}$ to $LBP_{P,R}^{riu2}$, which has $P+2$ distinct output values, is best implemented with a lookup table of 2^P elements.

$LBP_{P,R}^{riu2}$ is an effective texture descriptor with low computation complexity and desirable robustness against common image operations, which makes it very suitable for extracting texture features in building perceptual image hashing. In this paper, $LBP_{P,R}^{riu2}$ is exploited to extract texture features from the original image to construct a secondary image L, which is actually a stable texture feature matrix. Specifically, the pre-processed image I_1 is divided into non-overlapped blocks with a fixed size of $N \times N$. There are totally $(M/N) \times (M/N)$ sub-blocks. For each sub-block $C_{i,j}$, its $LBP_{P,R}^{riu2}$histogram is computed as follows to obtain the texture features.

$$I_1 = \begin{bmatrix} C_{1,1} & C_{1,2} & \cdots & C_{1,M/N} \\ C_{2,1} & C_{2,2} & \cdots & C_{2,M/N} \\ \cdots & \cdots & \cdots & \cdots \\ C_{M/N,1} & C_{M/N,2} & \cdots & C_{M/N,M/N} \end{bmatrix} \tag{11}$$

$$H(k) = \sum_{i=1}^{N} \sum_{j=1}^{N} C\left(LBP_{P,R}^{riu2}(i,j), k\right) \ k \in [0, K-1] \tag{12}$$

$$C(x,y) = \begin{cases} 1 & x = y \\ 0 & x < 0 \end{cases} \tag{13}$$

where K is the number of patterns, which is equal to $P+2$. The $LBP_{P,\ R}^{riu2}$ feature for each sub-block is a local histogram with $P+2$ bins. Then, the secondary image L, whose size is $(P+2 \times M/N \times M/N)$, is constructed as follows.

$$L_{i,j} = [H(1), H(2), \ldots H(10)] \tag{14}$$

$$L = \begin{bmatrix} L_{1,1} & L_{1,2} & \cdots & L_{1,M/N} \\ L_{2,1} & L_{2,2} & \cdots & L_{2,M/N} \\ \cdots & \cdots & \cdots & \cdots \\ L_{M/N,1} & L_{M/N,2} & \cdots & L_{M/N,M/N} \end{bmatrix} \tag{15}$$

4.3 NMF

NMF, which is also known as non–negative matrix approximation for multivariate data, is widely-accepted as an effective tool for the dimensionality reduction of large scale data. Different from the holistic representations by Principal Components Analysis (PCA), Vector Quantization (VQ) and Singular Value Decomposition(SVD), NMF finds a low rank by using non–negative decomposition of matrices with non–negative values, which make the non–negative data of matrix meaningful. NMF has been successfully used in various fields such as data clustering, image representation, image analysis and face recognition. Given a non–negative matrix3 $V = V(i,j)_{n \times m}$, NMF is to find non–negative matrix factors W and H such that:

$$V \approx W \times H \tag{16}$$

in which it meets $W = W(i,j)_{n \times r}$ and $H = H(i,j)_{r \times m}$, and r is usually chosen to be smaller than n or m. Thus, two matrices W and H are smaller than the original matrix V. Thus, the matrix V is approximately factorized into an $n \times r$ matrix W and an $r \times m$ matrix H. Please note that two matrices W and H are referenced as the base matrix and the coefficient matrix, respectively. That is, each data vector v is approximated by a linear combination of the columns of W, weighted by the components of h. Therefore, W can be regarded as containing a basis that is optimized for the linear approximation of the data in V. Since relatively few basis vectors are used to represent many data vectors, good approximation can only be achieved if the basis vectors discover the structure that is latent in the data.

All elements in W and H should be bigger than zero (non-negative) and r is the rank for NMF, $r < \min(n, m)$. Specifically, NMF usually uses the multiplicative update algorithms to find the non-negative matrix factors W and H as follows.

$$
\begin{cases}
W_{i,r} \leftarrow W_{i,r} \left(\dfrac{\sum_{j=1}^{n} H_{r,j} V_{i,j} / (WH)_{i,j}}{\sum_{j=1}^{n} H_{r,j}} \right) \\[6mm]
H_{r,j} \leftarrow W_{r,j} \left(\dfrac{\sum_{i=1}^{m} H_{i,r} V_{i,j} / (WH)_{i,j}}{\sum_{i=1}^{m} H_{r,j}} \right)
\end{cases}
\tag{17}
$$

$i = 1,2 \ ..., \ n; \ j = 1,2, \ ..., \ m; \ r = 1,2 \ ..., \ r$

The Kullback-Leibler (KL) divergence is used as the approximation criterion as follows:

$$
G = \sum_{i=1}^{n} \sum_{j=1}^{m} \left[V_{i,j} \log \frac{V_{i,j}}{(WH)_{i,j}} - V_{i,j} + (WH)_{i,j} \right]
\tag{18}
$$

To save storage space and transmission cost of image hash, NMF is applied to the secondary image L. The factorization factor $H = H(i,j)_{r \times m}$ is used to obtain the low rank matrix approximation to the secondary image L, which is referred to be the $LBP_{P, R}^{riu2}$-NMF feature matrix LN. The length of LN feature is depend on the length of $LBP_{P, R}^{riu2}$ feature, and the length of $LBP_{P, R}^{riu2}$ is $(M/N) \times (M/N)$, the rank of NMF is set as $r = 1$. Thus, the length of LN is $K = r \times (M/N) \times (M/N)$.

4.4 Color feature extraction

Color conveys important information for a color image, which makes a color image be much more vivid than a grayscale image. Since color images are much more popular nowadays than grayscale images, color information should also be involved for color image hashing. In this paper, the color feature $f_{x,y}^{color}$ is simply computed by averaging the red (R), green (G) and blue (B) components of each sub-block. However, the use of color feature greatly improves the discrimination capability, which will be illustrated in the experimental results.

$$f_{x,y}^{color} = \frac{\sum_{x=1}^{N}\sum_{y=1}^{N}(R(x,y)+G(x,y)+B(x,y))}{3} \tag{19}$$

$$F = \begin{bmatrix} f_{1,1}^{color} & f_{1,2}^{color} & \cdots & f_{1,M/N}^{color} \\ f_{2,1}^{color} & f_{2,1}^{color} & \cdots & f_{2,M/N}^{color} \\ \cdots & \cdots & \cdots & \cdots \\ f_{M/N,1}^{color} & f_{M/N,2}^{color} & \cdots & f_{M/N,M/N}^{color} \end{bmatrix} \tag{20}$$

4.5 Hash generation

Image hash function maps the input image into a binary hash string. After extracting the image content-related feature vectors from the input image, the feature vectors are required to be compressed and/or converted into a binary hash string. In this paper, the $LBP_{P,R}^{riu2}$-NMF feature LN and the color feature F are simply concatenated to form image hash, which is referred to $LBP_{P,R}^{riu2}$-NMF-Color hashing. First, LN is converted into a binary sequence as follows:

$$h_i^A = \begin{cases} 0 & if\, LN_i < LN_{i-1} \\ 1 & otherwise \end{cases} \tag{21}$$

where $0 \le i \le K$, $K = r \times (M/N \times M/N)$. Similarly, F is also quantized as follows:

$$h_i^B = \begin{cases} 0 & if\, F_j < F_{j-1} \\ 1 & otherwise \end{cases} \tag{22}$$

where $0 \le j \le Q$, $Q = (M/N \times M/N)$. Finally, the image hash h is obtained by simple concatenation of hA and hB as follows:

$$h = \left[h_1^A, h_2^A, \ldots, h_K^A, h_1^B, h_2^B, \ldots, h_Q^B \right] \tag{23}$$

Thus, the length of image hash is $T = K + Q$ bits. To further increase the security of the proposed image hashing approach, the numbers of K and Q are used to generate two secret keys $K1$ and $K2$, respectively. That is, the final image hash is obtained by hA and hB, which are controlled with $K1$ and $K2$.

$$h = \left[h_i^A \cdot K_i^1, h_j^B \cdot K_j^2 \right], \ 0 \le i \le K, \ 0 \le j \le Q \tag{24}$$

5. Experimental results and analysis

In the experiments, the proposed perceptual image hashing approach is implemented with Matlab 2016a. The hardware platform is as follows: a Lenovo PC with 2.70G Hz Intel Pentium G630 CPU, 2.0GB RAM and Windows 7 64-bit operating system. The size of 256×256 is set for image regularization. For the 3×3 Gaussian low-pass filter, it is defined with a mean of 0 and a standard deviation of 1. The normalized image I_1 is divided into non-overlapped blocks of size 32×32. The secondary image L is of size 10×64. For the $LBP_{P, R}^{riu2}$ operator, its neighborhood radius is 1, the sampling points are 8 and the rank of NMF is 1. Specifically, the related parameters are summarized as follows: $M = 256$; $N = 32$; $R = 1$; $P = 8$; $r = 1$; $K = 64$ and $Q = 64$. Thus, the final length of image hash is $T = K + Q = 128$ bits.

5.1 Hash similarity evaluation

Since our hash values are binary strings, the well-known Hamming distance is used as the distance metric to measure the similarity between two image hashes. Let H_1 and H_2 be two image hashes. The Hamming distance between H_1 and H_2 is defined as:

$$d_H(H_1, H_2) = \sum_{i=1}^{T} |H_1(i) - H_2(i)| \tag{25}$$

where $H_1(i)$ and $H_2(i)$ are the i-th elements of H_1 and H_2, respectively. If the Hamming distance is smaller than a pre-defined threshold, two images will be classified as a pair of similar images or visually identical images. Otherwise, they are considered to be different images with distinct contents.

5.2 Perceptual robustness

To evaluate the perceptual robustness of the proposed approach, 24 original images are selected from the Kodak Lossless True Color Image Suite [41] for experiments. And the well-known software tools, such as Matlab, Photoshop and StirMark 4.0 are used to generate visually similar versions of the test images. Actually, 10 content-preserving operations, which are considered as robustness attacks, are used to obtain visually identical versions of test images. Table 2 summarizes the detailed parameter settings for the used image operations. There are 74 visually similar images generated for each original image. Thus, the total number of test images will be $24 \times 74 + 24 = 1800$. After obtaining those visually similar images, the hamming distances are computed for each pair hashes between the original images and their similar versions. More intuitively, Fig. 3 shows the mean values of the first operation, and Table 3 reports the maximum, minimum, mean and standard deviation of the hamming distances for each content-preserving operation. It is apparent from Table 3 that the mean distances are smaller than 30, and the standard deviations are also small for them.

Table 2 Parameter settings for content-preserving image operations.

Operation	Parameters	Parameter Setting	Number
Brightness adjustment	Photoshop scale	$\pm 10, \pm 20$	4
Contrast adjustment	Photoshop scale	$\pm 10, \pm 20$	4
Gamma Correction	γ	0.7, 0.9, 1.1, 1.2	4
3×3 Gaussian low-pass filtering	Standard deviation	0.3, 0.4, ..., 1.0	8
Speckle noise	Variance	0.001, 0.002, ..., 1.010	10
Salt and pepper noise	Density	0.001, 0.002, ..., 1.010	10
JPEG compression	Quality factor	30, 40, ..., 100	8
PSNR	dB	10, 20, ..., 100	10
Image scaling	Ratio	0.5, 0.75, 0.9, 1.1, 1.5, 2	6
Rotation, cropping and rescaling	Agree in degree	1, 2, 3, ..., 10	10
Total			74

Source: author.

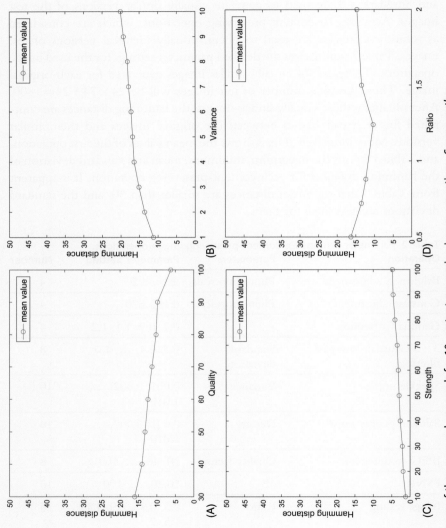

Fig. 3 Robustness of the proposed approach for 10 content-preserving image operations. *Source: author.*

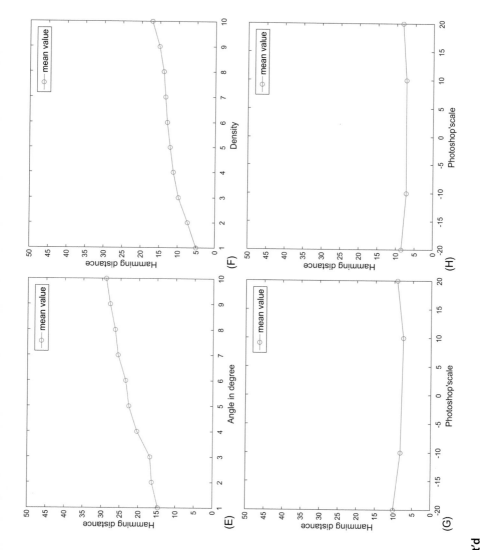

Fig. 3—Cont'd

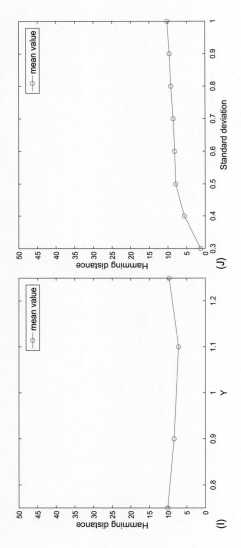

Fig. 3—Cont'd

Table 3 Statistics results of the hamming distance for 1776 pairs of visually similar images.

Operation	Maximum	Minimum	Mean	Standard Deviation
Brightness adjustment	19	0	8.21	0.58
Contrast adjustment	15	0	7.62	0.10
Gamma Correction	18	0	8.35	0.45
3×3 Gaussian low-pass filtering	26	0	7.98	1.54
Speckle noise	29	0	16.40	0.70
Salt and pepper noise	28	0	11.35	0.55
JPEG compression	25	0	11.55	0.64
PSNR	21	0	2.74	0.62
Image scaling	29	0	12.61	1.08
Rotation, cropping and rescaling	30	0	20.49	1.61

Source: author.

5.3 Discriminative capability

To validate the discriminative capability of the proposed approach, another 1338 color images are selected from the true color images database (UCID) [42] for experiments. These images contain quite diverse contents such as human beings, landscape, buildings and sport. First, the image hashes are computed for these 1338 original images, respectively. Then, the hamming distances of image hashes are computed between every color image and the rest 1337 color images. Thus, there are totally $1338 \times (1338-1)/2 = 894,453$ results, which are available for validating the discriminative capability of the proposed approach. The minimum, maximum, mean and standard deviation of the hamming distances among the 1338 images from the UCID dataset are 17, 100, 63.21 and 6.83, respectively. Further, to identify the most probable distribution among the 894,453 results, the well-known hypothesis testing method, i.e., Chi-square test [43] is adopted to examine the Poisson distribution, the Rayleigh distribution, the Weibull distribution, the lognormal distribution, the normal distribution and the Gamma distribution. The parameters of these distributions are first estimated by maximum likelihood estimation. Then, to measure the differences between the observed values and the theoretical values, their statistics χ^2 are calculated as follows:

$$\chi^2 = \sum_{i=0}^{N} \frac{(n_i - q_i)^2}{q_i} \tag{26}$$

$$q_i = n_{total} P_i \tag{27}$$

where n_i and q_i are the i-th hamming distance frequency of the observed value and the theoretical value, respectively. n_{total} is the number of trials, N is the hash length, and P_i is the probability at i determined by the probability density function of the candidate distribution to be evaluated. Table 4 summarizes the estimated parameters and the chi-square results of the–above mentioned candidate distributions. From it, we found that the smallest one is the χ^2 result of Normal distribution. The mean value and the standard deviation of Normal distribution are 63.23 and 6.85, respectively, which is highly similar with the empirical distribution. Thus, Normal distribution is the most fitting distribution to model the hamming distances among the 1338 images. In Fig. 4, comparison is made between the actual/empirical distribution of hamming distances and the theoretical/ideal normal distribution. We can observe that the majority of hamming distances is around 63. Moreover, the hamming distances for different images are much bigger than those for visually identical images. Thus, the proposed approach achieves desirable discriminative capability.

Further, we assess the probability of false detection, which means that different images wrongly detected as similar images. For a given threshold T, the collision probability P of false detection can be calculated by the following equation.

$$P(d_H \le T) = \frac{1}{\sqrt{2\pi}\sigma} \int_0^T \exp\left[-\frac{(\chi - \mu^2)}{2\sigma^2}\right] dx \\ = \frac{1}{2} erfc\left(-\frac{T-\mu}{\sqrt{2}\sigma}\right) \tag{28}$$

Table 4 The estimated parameters and χ^2 results of the test distributions.

Distribution	Estimated parameters	χ^2
Normal	$\mu=63.23$, $\sigma=6.85$	282,577
Poisson	$\lambda=63.23$	4.51×10^5
Lognormal	$\mu=4.14$, $\sigma=0.11$	1.88×10^{25}
Rayleigh	$\beta=44.97$	2.19×10^6
Gamma	$a=83.66$, $b=0.76$	7.79×10^{15}
Weibull	$\beta=66.29$, $\eta=9.87$	9.72×10^{18}

Source: author.

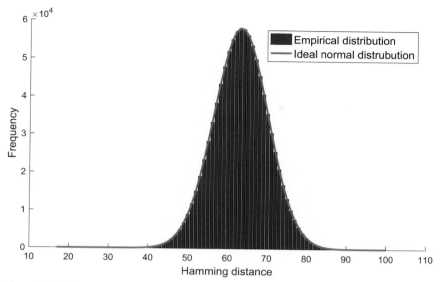

Fig. 4 Hash distance distribution between different images. *Source: author.*

Table 5 Collision probabilities under different thresholds.

Threshold	Collision probability
18	1.96×10^{-11}
21	3.44×10^{-10}
24	5.01×10^{-9}
27	6.04×10^{-8}
30	6.05×10^{-7}
33	5.03×10^{-6}

Source: author.

where erfc( ) is the complementary error function. Table 5 reports the collision probabilities of false detection under different thresholds. Apparently, the smaller the threshold T is set, the smaller collision probability P is obtained. That is, a small T means a low probability of false detection, i.e., a good discrimination. However, the hamming distance between the hash pair of two visually similar images should be smaller than the threshold T. A small threshold T also inevitably reduces the robustness. In practice, a proper threshold T should be selected to compromise well between perceptual robustness and discrimination capability. From Fig. 4, the hamming distance of the original image and its visually similar versions by common content-preserving operations are almost all smaller than 30. When T is equal to 30, our proposed approach achieves the collision probability of 6.05×10^{-7}, which is small enough. Thus, we set T with 30 to compromise between perceptual robustness and anti-collision capability.

5.4 Key dependence

Besides the perceptual robustness and discrimination capability, key dependence is also important for perceptual image hashing. That is, an image hashing should satisfy that different secret keys produce significantly different image hashes. As described in Section 4.5, two secret keys $K1$ and $K2$ are used to obtain more secure image hashing. In the experiments, the color images from the UCID dataset are also used as test images. Fig. 5 reports the key dependence performance of the proposed approach, in which the X-axis represents 100 groups of randomly generated keys $K1$ and $K2$, and the Y-axis is the hamming distance between the hashes controlled by different groups of keys. From it, we can observe that the minimum hamming

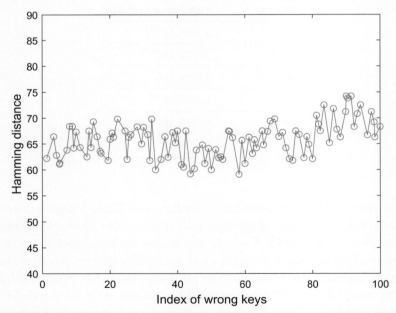

Fig. 5 Hamming distance between hashes of an image controlled with 100 group keys.
Source: author.

distance is 53, and especially all hamming distances are much bigger than $T=30$. This implies that it is difficult for malicious users to generate an image with the same hash, if they have no correct keys. That is, the proposed image hashing approach is key-dependent, which satisfies the security requirement.

5.5 Performance comparisons

In this sector, performance comparisons are made between the proposed approach and existing image hashing schemes. Specifically, five typical existing works, which include the SVD-CSLBP hashing [20], the Random Walk hashing [44], the Multi- Histogram hashing [45], the MDS hashing [46] and the CVA-Canny hashing [47], are selected as the benchmarks. Thus, there are totally six image hashing approaches including the proposed approach for performance comparisons. Further, we set these existing schemes with the same parameters that they were reported, thus fair comparisons are made. For the SVD-CSLBP hashing, input images are first converted into gray-scale images and then resized to 256×256. For the Random Walk hashing, the input image is divided into the blocks of size 20×20, $n=20$ and $T=1.5$. For the Multi-Histogram hashing, the input image is resized to 512×512, the ring number is $n=4$, and the segment number is b$=4$. For the MDS hashing, the image is resized to 256×256, and the value of d is set with 40. For the CVA-Canny hashing, K is set with 25 and $\Delta d=3$.

Fig. 6 compares the visual classification results among the six image hashing approaches, in which the blue and purple represent the similarity results for those visually identical images and different images, respectively. For each image hashing approach, there always exist some overlapping regions of the distribution results between visually identical images and different images. For an image hashing approach with excellent visual classification performance, the overlapping regions of visually identical and different images should be small enough. In the experiments, the number of visually identical images and different images in the overlapping intervals are [0.74456, 2], [0, 0.51042], [4, 183707], [17, 30], [0.6512, 1] and [−0.1381, 1] for the SVD-CSLBP hashing [20], the Random walk hashing [44], the Multi-Histogram hashing [45], the MDS hashing [46], the CVA-Canny hashing [47] and the proposed approach, respectively. To facilitate the experimental comparisons, the overlapping regions are normalized. The percentages of overlapping regions are 0.62772, 0.51042,

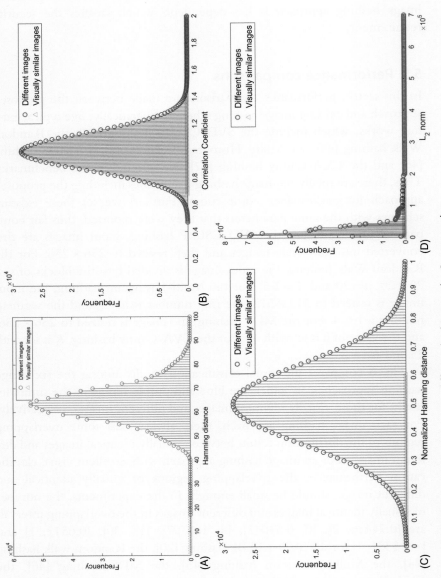

Fig. 6 Visual classification comparisons among different hashing works. *Source: author.*

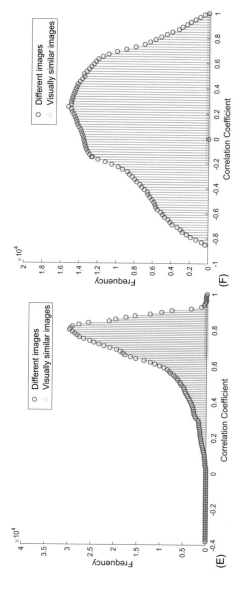

Fig. 3—Cont'd

0.27645, 0.3488, 0.5842 and 0.13 for the SVD-CSLBP hashing, the Random walk hashing, the Multi-Histogram hashing, the MDS hashing, the CVA-Canny hashing and the proposed approach, respectively. That is, the proposed approach has the smallest percentage of overlapping regions among six image hashing approaches. Thus, the proposed approach achieves much better classification performance than the rest five existing approaches. Moreover, the receiver operating characteristics (ROC) curve [48], which is well-known in describing the classification and discrimination capability, is also used to analyze the experimental results. The ROC curve is created by plotting the true positive rate (P_{TPR}) against the false positive rate (P_{FPR}). Specifically, the X-axis and the Y-axis represent P_{FPR} and P_{TPR} in the ROC curve, respectively. The definitions of P_{FPR} and P_{TPR} are as follows.

$$P_{FPR}(\rho \leq T) = \frac{N_{false}}{N_1} \tag{29}$$

$$P_{TPR}(\rho \leq T) = \frac{N_{true}}{N_2} \tag{30}$$

where N_{false} is the number of distinct images which are wrongly considered as visually similar images, and N_{true} is the number of visually identical images which are correctly considered as similar images, N_1 is the total pairs of different images, N_2 is the total pairs of visually identical images. Apparently, P_{FPR} and P_{TPR} can represent the perceptual robustness and discriminative capability, respectively.

For five existing image hashing approaches, we directly use the same thresholds that were reported in their original works, so as to obtain their P_{FPR} and P_{TPR} results. Fig. 7 reports the ROC curves of five existing works and the proposed approach. From it, the proposed approach achieves a ROC curve which is the closest to the top left corner, whereas the $LBP_{P,R}^{riu2}$-NMF achieves a ROC curve that is the second closest to the top left corner. Moreover, we can observe from the ROC curves that when the FPR is close to 0, the TPRs of the SVD-CSLBP hashing, the Random Walk hashing, the Multi-Histogram hashing, the MDS hashing, the CVA-Canny hashing, the $LBP_{P,R}^{riu2}$-NMF hashing, and the proposed $LBP_{P,R}^{riu2}$-NMF-Color approach are 0.6194, 0.665, 0.705, 0.8756, 0.8699, 0.9018 and 0.9454, respectively. When the TPR reaches 1, their FPRs are 0.7906, 0.4722, 0.6182, 0.3266, 0.3759, 0.0564 and 0.0225, respectively. Further, the area under the curve (AUC) (Fawcett T. 2006) is also computed to compare the performance. Table 6 reports the comparison results of AUC, average running time and

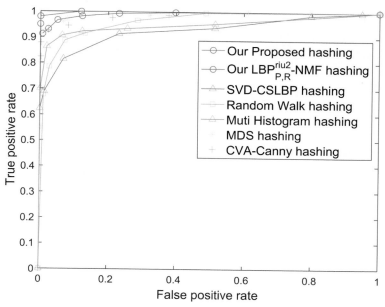

Fig. 7 Comparison of the ROC curves among different hashing works. *Source: author.*

Table 6 Performance comparisons among different image hashing works.

Performance	AUC	Average running times(s)	Hash length
SVD-CSLBP hashing [20]	0.9231	0.26	64 decimal
Random Walk hashing [44]	0.9559	0.08	16 decimal
Multi-Histogram hashing [45]	0.9414	0.09	144 decimal
MDS hashing [36]	0.9869	0.45	180 decimal
CVA-Canny hashing [47]	0.9810	0.36	40 integers
$LBP^{riu2}_{P,\,R}$-NMF hashing	0.9809	0.13	64 bits
Proposed hashing	0.9954	0.16	128 bits

Source: author.

hash length. From it, the $LBP_{P,\,R}{}^{riu2}$-NMF based image hashing achieves much better performances than existing works in terms of both perceptual robustness and discrimination capability. By further exploiting the color features, the proposed $LBP_{P,\,R}{}^{riu2}$-NMF–Color approach achieves the best discrimination capability. For the average running time, the proposed approach consumes more time than the Multi-Histogram hashing [45], the Random Walk hashing [44] and $LBP_{P,\,R}^{\,riu2}$-NMF hashing, but it is much faster than the SVD-CSLBP hashing [20], the MDS hashing [46] and the CVA-Canny hashing [47]. The hash length of the proposed approach is 128 bits, which is also suitable for storage and transmission.

5.6 Summary of the lessons learned from this research

Based on our experience developing the perceptual image hashing scheme presented in this chapter, we learned the following lessons:

First, feature extraction and hash generation are the key issues for the design of any perceptual image hashing scheme. In the feature extraction stage, the perceptual image hashing system extracts some image features from the input image to generate the continuous hash vector. In the hash generation stage, the continuous hash vector is quantized into the discrete hash vector, and then the discrete hash vector is converted into the binary perceptual hash string. Finally, the binary perceptual hash string is compressed and encrypted into the final hash which is usually compact. That is, hash generation can consist of quantization, compression and encryption. However, if the binary perceptual hash string is shorter enough by selecting appropriate features, the steps of both compression and encryption can be omitted. Moreover, if there are too many features extracted to enhance the discriminative capability of perceptual image hashing, data reduction technique might be exploited before quantization. Specifically,

Second, NMF, which is also known as non–negative matrix approximation, is an effective data reduction technique for data clustering. NMF has shown better performance than principal components analysis (PCA) and vector quantization (VQ) in learning parts-based representation. For perceptual image hashing, NMF can be exploited as an efficient technique for the dimensionality reduction of feature vectors. That is, a high-dimensional feature matrix can be approximately factorized into two matrixes whose dimensions are much smaller than the original feature matrix. Thus, NMF can effectively achieve the compactness requirement of perceptual image hashing.

Third, since color conveys important information, which makes a color image be much more vivid than a grayscale image. Thus, color information should also be involved in the design of a perceptual hashing scheme for color images. Moreover, the use of color features greatly improves the discrimination capability of perceptual image hashing.

6. Conclusion

In essence, perceptual image hashing maps the input image into a short sequence that represents the principle contents of the image, which can play an increasingly important role in digital forensics, near duplicate image detection and source tracking of the massive image information. In this paper, a robust perceptual image hashing approach is proposed by exploiting the $LBP_{P,R}^{riu2}$-NMF-color features. To guarantee the perceptual robustness against content-preserving operations and reduce the length of image hash, Gaussian low-pass filtering and NMF are first applied on the secondary image of $LBP_{P,R}^{riu2}$. Then, the color feature of each R, G and B sub-block is extracted to further improve the discriminative capability. The ROC curves show that the proposed perceptual image hashing approach achieves satisfactory perceptual robustness and discriminative capability simultaneously. However, the proposed approach is still not robust to large angle rotation. For future research, we will further investigate the design of perceptual image hashing with any-angle rotation robustness and stronger discriminative capability. In addition, we will further investigate the application of perceptual image hashing for cloud-based encrypted multimedia system, in which the security and privacy threats of the image data in cloud servers also poses great challenges.

Acknowledgment

This work is supported in part the National Natural Science Foundation of China (61972143).

Key Terminology and Definition

Cryptographic hash function A cryptographic hash function is a special class of hash function that has certain properties including pre-image resistance, second pre-image resistance and collision resistance, which makes it suitable for use in cryptography. In essence, a cryptographic hash function is a mathematical algorithm that maps data of arbitrary size to a bit string of a fixed size (a hash). The values returned by a cryptographic hash function are called message digest or simply hash values. A cryptographic hash function is usually designed to be a one-way function. That is, it is a function that is infeasible to invert. In cryptography, cryptographic hash functions can be divided into two main categories. In the first category are those functions whose designs are based on a mathematical problem and thus their security follows from rigorous mathematical proofs, complexity theory and formal reduction. In the second category are functions that are not based on mathematical problems but on an Ad hoc basis, where the bits of the message are mixed to produce the hash. They are then believed to be hard to break, but no such formal proof is given. Commonly used cryptographic hash functions include MD5 and SHA-1, although many others also exist.

Perceptual image hashing Different from the cryptographic hash functions to authenticate image data, perceptual image hashing has the requirement of perceptual robustness or tamper localization. Perceptual Image hashing extracts certain features from image data, and then calculates a hash value based on the extracted features. By comparing the hash values of the original image and the image to be authenticated, the authentication and integrity of image content are achieved. It is expected that a perceptual image hash scheme should survive on those acceptable content-preserving manipulations, while rejecting those malicious or content-changing manipulations. In general, a perceptual image hash function must meet the following five properties: perceptual robustness, uniqueness, unpredictability, pair-wise independence and compactness. Since these five requirements are conflicting, it is extremely difficult for an adversary to modify the essential content of an input image yet keep the hash value unchanged. Especially, trade-offs must be made between perceptual robustness and discriminative capability, which is usually the key issue to be considered for the design of any perceptual image hash function.

Image forensics Image forensics aims at providing powerful tools to support blind investigation about the authenticity and integrity of image data, which is quite different from existing multimedia security techniques such as watermarking and steganography. By exploiting image processing and analysis tools, image forensics attempts to recover information about the history of an image. Existing image forensics techniques can be divided into two categories. The first category is image source device identification, which identifies the device that captured the image, or at least determines which devices did not capture it. The second category is image forgery detection, which exposes the traces of image manipulation (i.e., forgeries) by studying any inconsistencies in natural image statistics.

Local Binary Pattern (LBP) LBP is a type of visual texture descriptor, which is usually used for classification in computer vision. The basic idea is to summarize the local structure in an image by comparing each pixel with its neighborhood. Take a pixel as the center and threshold its neighboring pixels. If the intensity of the center pixel is greater than or equal with its neighbor, the neighboring pixel will be denoted t with 1, or else 0. Then, a binary number just like 11001111 will be obtained for each pixel. For a center pixel with 8 surrounding pixels, there are totally $2^8 = 256$ possible combinations. Actually, this is the simplest form of Local Binary Patterns (LBP), which is sometimes abbreviated as *LBP codes*. There are many extensions to the basic LBP to improve its performance, which include Uniform Local Binary Patterns, Over-Complete Local Binary Patterns (OCLBP), Transition Local Binary Patterns (tLBP), Direction coded Local Binary Patterns (dLBP), Modified Local Binary Patterns (mLBP) and so on.

Rotation Invariant Uniform Local Binary Patterns Among various extensions to the basic LBP, Rotational Invariant Uniform Local Binary Patterns are designed for the rotation invariant texture analysis on 3D volume data. The term "uniform" refers to the uniform appearance of the local binary patterns. That is, there is only a limited number of transitions or discontinuities in the circular presentation of the patterns. For example, a local binary pattern is called uniform if the binary pattern contains at most two bitwise transitions from 0 to 1 or vice versa when the bit pattern is considered circular. In the computation of the LBP histogram, uniform patterns are used so that the histogram has a separate bin for every uniform pattern and all non-uniform patterns are often assigned to a single bin.

Non-negative Matrix Factorization (NMF) NMF is also known as non-negative matrix approximation, which is actually a group of algorithms in multivariate analysis and linear algebra where a matrix V is factorized into (usually) two matrices W and H, with the property that all three matrices have no negative elements. This property of non-negativity is a useful constraint for matrix factorization that can learn a parts representation of the data, which makes the resulting matrices easier to inspect. Specifically, NMF can be applied to the statistical analysis of multivariate data in the following manner. Given a set of multivariate n-dimensional data vectors, the vectors are placed in the columns of an $n \times m$ matrix V where m is the number of examples in the data set. This matrix is then approximately factorized into an $n \times r$ matrix W and an $r \times m$ matrix H. Usually r is chosen to be smaller than m or n, so that W and H are smaller than the original matrix V. This results in a compressed version of the original data matrix.

References

[1] P. Singh, B. Raman, N. Agarwal, P.K. Atrey, Secure cloud-based image tampering detection and localization using POB number system, ACM Trans. Multimed. Comput. Commun. Appl. 13 (3) (2017) 230–239.

[2] Z.S. Liang, G.B. Yang, X.L. Ding, L.D. Li, An efficient forgery detection algorithm for object removal by exemplar-based image inpainting, J. Vis. Commun. Image Represent. 30 (2015) 75–85.

[3] X.L. Ding, G.B. Yang, R. Li, L.B. Zhang, Y. Li, X.M. Sun, Identification of MC-FRUC based on spatial-temporal Markov features of residue signal, IEEE Trans. Circuits Syst Video Technol. 28 (7) (2018) 1497–1512.

[4] T.Y. Lee, S.F. Lin, Dual watermark for image tamper detection and recovery, Pattern Recognit. 41 (11) (2008) 3497–3506.

[5] F. Ahmed, M.Y. Siyal, V.U. Abbas, A secure and robust hash-based scheme for image authentication, Signal Process. 90 (5) (2010) 1456–1470.

[6] A. Tiwaria, M. Sharmaa, R.K. Tamrakar, Watermarking based image authentication and tamper detection algorithm using vector quantization approach, AEU-Int. J. Electron. C. 78 (2017) 114–123.

[7] D.Y. Zhang, T. Yin, G.B. Yang, L.D. Li, X.M. Sun, Detecting image seam carving with low scaling ratio using multi-scale spatial and spectral entropies', J. Vis. Commun. Image Represent. 48 (2017) 281–291.

[8] H. Azhar, P. William, A.E.S. Brahim, A.O. Abdellah, Perceptual Image Hashing, in: M.D. Gupta (Ed.), Watermarking-Volume 2, InTech Press, 2012, ISBN: 978-953-51-0619-7. Retrieved June 10, 2016, from http://cdn.intechopen.com/pdfs/36921/InTech-Perceptual_image_hashing.pdf.

[9] J. Singh, K. Lata, J. Ashraf, Image encryption & decryption with symmetric key cryptography using MATLAB, Int. J. Curr. Eng. Technol. 5 (1) (2015) 448–451.

[10] K. Wang, J. Tang, N. Wang, L. Shao, Semantic boosting cross-modal hashing for efficient multi-media retrieval, Inform. Sci. 330 (2016) 199–210.

[11] C. Winter, M. Steinebach, Y. Yannikos, Fast indexing strategies for robust image hashes, Digital Investig. 11 (2014) 27–35.

[12] F.H. Zou, H. Feng, H.F. Ling, C. Liu, L.Y. Yan, P. Li, Nonnegative sparse coding induced hashing for image copy detection', Neurocomputing 105 (1) (2013) 81–89.

[13] C.P. Yan, C.M. Pun, X.C. Yuan, Multi-scale image hashing using adaptive local feature extraction for robust tampering detection', Signal Process. 121 (2016) 1–16.

[14] X. Lv, Z.J. Wang, Reduced-reference image quality assessment based on perceptual image hashing, in: IEEE International Conference on Image Processing (ICIP). IEEE. Cairo, Egypt, 2009, pp. 4361–4364.

[15] J.Q. Yang, J. Xie, G.P. Zhu, S. Kwong, Y.Q. Shi, An effective method for detecting double JPEG compression with the same quantization matrix, IEEE Trans. Inf. Forensics Secur. 9 (11) (2014) 1933–1942.

[16] G. Cao, Y. Zhao, R.R. Ni, X.L. Li, Contrast enhancement-based forensics in digital images, IEEE Trans. Inf. Forensics Secur. 9 (3) (2014) 515–525.

[17] K. Wattanachote, T.K. Shih, W.L. Chang, H.H. Chang, Tamper detection of JPEG image due to seam modifications, IEEE Trans. Inf. Forensics Secur. 10 (12) (2015) 2477–2491.

[18] T. Yin, G.B. Yang, L.D. Li, D.Y. Zhang, Detecting seam carving based image resizing using local binary patterns, Comput. Secur. 55 (2015) 130–141.

[19] D. Cozzolino, G. Poggi, L. Verdoliva, Efficient dense-field copy–move forgery detection, IEEE Trans. Inf. Forensics Secur. 10 (11) (2015) 2284–2297.

[20] R. Davarzani, S. Mozaffari, K. Yaghmaie, Perceptual image hashing using center-symmetric local binary patterns, Multimed. Tools Appl. 75 (8) (2016) 4639–4667.

[21] J. Fridrich, M. Goljan, Robust hash functions for digital watermarking', in: International Conference on Information Technology: Coding and Computing, Las Vegas, USA, 2000, pp. 178–183.

[22] C.Y. Lin, S.F. Chang, A robust image authentication method distinguishing JPEG compression from malicious manipulation, IEEE Trans. Circuits Syst Video Technol. 11 (2) (2001) 153–168.

[23] B. Zhang, X. Yang, X.X. Niu, K.G. Yuan, An anti-JPEG compression image perceptual hashing algorithm, in: International Conference Applied Informatics and Communication (ICAIC). Xi'an, China, 2011, pp. 397–406.

[24] L.J. Yu, S.H. Sun, Image robust hashing based on DCT sign, in: International Conference on Intelligent Information Hiding and Multimedia(IIH-MMSP). Pasadena, CA, USA, 2006, pp. 131–134.

[25] R. Venkatesan, S.M. Koon, M.H. Jakubowski, P. Moulin, Robust image hashing', in: IEEE International Conference on Image Processing (ICIP), vol. 3, IEEE, Vancouver, Canada, 2000, pp. 664–666.

[26] A. Swaminathan, Y. Mao, M. Wu, Robust and secure image hashing', IEEE Trans. Inf. Forensics Secur. 1 (2) (2006) 215–230.

[27] P. Supakorn, K. Fouad, B. Ahmed, Fourier-Mellin transform for robust image hashing, in: Fourth International Conference on Emerging Security Technologies. Cambridge, UK, 2013, pp. 58–61.

[28] R. Sun, W.J. Zeng, Secure and robust image hashing via compressive sensing, Multimed. Tools Appl. 70 (3) (2014) 1651–1665.

[29] Y.Q. Lei, Y.G. Wang, J.W. Huang, Robust image hash in radon transform domain for authentication', Signal Process. Image Commun. 26 (6) (2011) 280–288.

[30] C.D. Roover, C.D. Vleeschouwer, F. Lefebvre, B. Macq, Robust image hashing based on radial variance of pixels', in: IEEE International Conference on Image Processing. Genova, Italy, 2005, pp. 50–53.

[31] X.C. Guo, D. Hatzinakos, Content based image hashing via wavelet and radon transform, in: Pacific-Rim Conference on Multimedia (PCM). Hong Kong, China, 2007, pp. 755–764.

[32] S.S. Kozat, R. Venkatesan, M.K. Mihak, Robust perceptual image hashing via matrix invariants, in: IEEE International Conference on Image Processing (ICIP), IEEE, Singapore, 2004, pp. 3443–3446.

[33] G. Lahouari, Robust perceptual colour image hashing using quaternion singular value decomposition', in: IEEE International Conference on Acoustics, Speech and Signal Processing (ICASSP). Florence, Italy, 2014, pp. 3794–3798.

[34] V. Monga, M.K. Mhcak, Robust and secure image hashing via non-negative matrix factorizations, IEEE Trans. Inf. Forensics Secur. 2 (3) (2007) 376–390.

[35] Z.J. Tang, S.Z. Wang, X.P. Zhang, W.M. Wei, Y. Zhao, Lexicographical framework for image hashing with implementation based on DCT and NMF, Multimed. Tools Appl. 52 (2) (2011) 325–345.

[36] Z.J. Tang, X.P. Zhang, S.Z. Wang, Robust perceptual image hashing based on ring partition and NMF, IEEE Trans. Knowl. Data Eng. 26 (3) (2014) 711–724.

[37] F. Khelifi, J.M. Jiang, Analysis of the security of perceptual image hashing based on non-negative matrix factorization', IEEE Signal Process. Lett. 17 (1) (2010) 43–46.

[38] L. Liu, M.Y. Yu, L. Shao, Unsupervised local feature hashing for image similarity search, IEEE Trans. Cybern. 46 (11) (2016) 2548–2558.

[39] X.D. Lv, Z.J. Wang, Perceptual image hashing based on shape contexts and local feature points, IEEE Trans. Inf. Forensics Secur. 7 (3) (2012) 1081–1093.

[40] R.K. Karsh, R.H. Laskar, B.B. Richhariya, Robust image hashing using ring partition-PGNMF and local features, Springerplus 5 (1) (2016) 1995–2015.

[41] Kodak. (2010) Lossless True Colour Image Suite image database. Retrieved March 15, 2017, from: www.r0k.us/graphics/kodak/.

[42] UCID. (2010) Nottingham Trent University UCID image database. Retrieved March 15, 2017, from http://vision.doc. ntu.ac.uk/datasets/UCID/ucid.html.

[43] S. Brandt, Testing statistical hypotheses, Publ. Am. Stat. Assoc. 101 (474) (2005) 847–848.

[44] X. Huang, X.G. Liu, G. Wang, M. Su, A Robust Image Hashing with Enhanced Randomness by Using Random Walk on Zigzag Blocking, IEEE Trustcom-BigDataSE-ISPA. IEEE, Tianjin, China, 2016, pp. 14–18.

[45] Z.J. Tang, L.Y. Huang, Y.M. Dai, F. Yang, Robust image hashing based on multiple histograms, Int. J. Digit. Content Technol. its Appl. 6 (23) (2012) 39–47.

[46] Z.J. Tang, Z.Q. Huang, X.Q. Zhang, H. Lao, Robust image hashing with multi-dimensional scaling, Signal Process. 137 (2017) 240–250.

[47] Z.J. Tang, Z.Q. Huang, X.Q. Zhang, H. Lao, Robust image hashing based on colour vector angle and Canny operator, AEU-Int. J. Electron. Commun. 70 (6) (2016) 833–841.

[48] T. Fawcett, An introduction to ROC analysis, Pattern Recognit. Lett. 27 (8) (2006) 861–874.

About the authors

Mr. Ming Xia: Dr. Ming Xia is currently a Ph.D. candidate at the School of Computer Science and Electrical Engineering, Hunan University, Changsha, China. He is also a lecture in the Southwest University for Nationalities, Chendu, China. His research areas are perceptual image hashing and passive image/video forensics.

Miss. Siwei Li: Miss Siwei Li was a Master student at the School of Computer Science and Electrical Engineering, Hunan University, Changsha, China. She obtained her Master degree in June 2018. Her research work focus on active image security, especially perceptual image hashing.

Mr. Weibing Chen: Mr. Weibing Chen is a professor at the School of Electronic Communication & Electrical Engineering, Changsha University, China. He obtained his Bachelor and Master degree from National University of Defense Technology (NUDT), China in 1990 and 2002, respectively. He worked in Hunan University, China from May 2005 to July 2010. He is also the Principle Investigator of several projects such as Hunan Provincial Science & Technology Project. His current research interests include mobile communication, multimedia communication and image content security. He has authored more than 40 papers in journals and conference proceedings.

Dr. Gaobo Yang: Dr. Gaobo Yang is a professor at the School of Computer Science and Electrical Engineering, Hunan University, Changsha, China. He is a key member in the Hunan Provincial Key Laboratory of Networks and Information Security, China. He obtained his Ph.D. degree from Shanghai University, China in 2004. From August 2010 to July 2011, he was a visiting scholar in University of Surrey, U.K. He is also the Principle Investigator of several projects, such as the Natural Science Foundation of China (NSFC), the Special Pro-phase Project on National Basic Research Program of China (973), and the Program for New Century Excellent Talents in University. His research areas include image and video signal processing, and digital media forensics. He has published 5 books and 100 + papers. He has mentored over 40 Ph.D. and Masters students.

Printed in the United States
by Baker & Taylor Publisher Services